Robotics and Mechatronics for Agriculture

Robotics and Mechatronics for Agriculture

Editors

Dan Zhang and Bin Wei
Department of Mechanical Engineering
Lassonde School of Engineering
York University
Toronto, Ontario, Canada

CRC Press
Taylor & Francis Group
Boca Raton London New York

CRC Press is an imprint of the
Taylor & Francis Group, an **informa** business

A SCIENCE PUBLISHERS BOOK

CRC Press
Taylor & Francis Group
6000 Broken Sound Parkway NW, Suite 300
Boca Raton, FL 33487-2742

First issued in paperback 2021

© 2018 by Taylor & Francis Group, LLC
CRC Press is an imprint of Taylor & Francis Group, an Informa business

No claim to original U.S. Government works

Version Date: 20170721

ISBN-13: 978-0-367-78172-9 (pbk)
ISBN-13: 978-1-138-70240-0 (hbk)

Library of Congress Cataloging-in-Publication Data

Names: Zhang, Dan, 1964- editor. | Wei, Bin, 1987- editor.
Title: Robotics and mechatronics for agriculture / editors: Dan Zhang and Bin Wei.
Description: Boca Raton, FL : CRC Press, Taylor & Francis Group, 2017. | Includes bibliographical references and index.
Identifiers: LCCN 2017033064 | ISBN 9781138702400 (hardback : alk. paper)
Subjects: LCSH: Robotics. | Mechatronics. | Agricultural innovations.
Classification: LCC S678.65 .R62 2017 | DDC 631.30285/63--dc23
LC record available at https://lccn.loc.gov/2017033064

Visit the Taylor & Francis Web site at
http://www.taylorandfrancis.com

and the CRC Press Web site at
http://www.crcpress.com

Preface

Robotics and mechatronics have been used in many arenas, one of which is the agricultural industry. Using robotic based machines in agriculture will become common in the future. Automatic machines will replace human beings in agriculture and they can greatly help farmers to achieve efficient farming. This book will focus on the robotics and mechatronics that are used in agriculture. The aim of the book is to introduce the state-of-the-art technologies in the field of robotics and mechatronics for agriculture in order to further summarize and improve the methodologies on the agricultural robotics. Advances made in the past decades have been described in this book.

We would like to thank all the authors for their contributions to the book. We are also grateful to the publisher for supporting this project and Vijay Primlani for his assistance both with the publishing venture and the editorial details. We hope the readers find this book informative and useful.

This book consists of 8 chapters. Chapter 1 focuses on the function and mechanism of aeration for process optimization. Chapter 2 discusses key aspects of a design of a robotic platform for the management of crops in agriculture. In particular, the system considered seeks to address the increasing threat of weed species resistant to herbicide. Chapter 3 presents a case study of an automated "field scout" ground platform equipped with the means for both sensing and manipulating its changing environment for the purpose of providing actionable data (including samples of physical field specimens) to a farmer. Chapter 4 presents a critical and detailed review about the application of simple color cameras to cover different aspects of agricultural industry. Chapter 5 presents some existing robotic based farming machineries, and some main issues in the robotic based farming are also illustrated. Chapter 6 reviews collaborative multi-agent systems in agricultural applications involving a RA to RA and RA to human agent (HA) collaboration. Common systems' control architecture and design, tools and middleware, planning and decision execution, cooperation behaviour, and communication systems are discussed with recently developed systems for agricultural applications. Chapter 7 proposes an adaptive and robust model predictive controller to address the problem of wheel slip in field vehicles. Chapter 8 focuses on model reference adaptive control of dynamical systems with matched system uncertainties but unmatched disturbances. The proposed control framework has a high potential to guarantee the completion of autonomous seeding, harvesting, and/or row cropping via unmanned ground vehicles, or farm imaging and monitoring via unmanned aerial vehicles with high accuracy.

Finally, the editors would like to acknowledge all the friends and colleagues who have contributed to this book.

Toronto, Ontario, Canada   **Dan Zhang**
February 2017   **Bin Wei**

Contents

1

Process Optimization of Composting Systems

Naoto Shimizu

1. INTRODUCTION TO PROCESS OPTIMIZATION OF COMPOSTING SYSTEMS

Food supply is a primary issue for people around the world. Increasing demand for food has been anticipated by the increased intake of meat, fat, processed foods, sugar and salt nutrition transition. The livestock (cattle, swine, chicken) sector is a substantial source of nutrients for human consumption. In Japan, total production of animal waste in 2015 was 83 million tons. There is a need to develop management systems that use cattle manure effectively and without causing adverse environmental effect.

Problems associated with waste from animal husbandry are, safety, financial and environmental. Huge amounts of solid wastes from animal husbandry result in odor problems that can lead to complaints from neighbors and other people. Composting is a simple and energy efficient way to solve this problem. The purposes of composting are:

- Elimination of pathogens and weeds
- Microbial stabilization
- Reduction of volume and moisture
- Removal and control of odors
- Ease of storage, transport and use

Field Science Center for Northern Biosphere, Research Faculty of Agriculture, Hokkaido University, Kita 9 Nishi 9, Kita-ku, Sapporo, Hokkaido 060-8589, Japan.
Email: shimizu@bpc.agr.hokudai.ac.jp

Many studies have addressed the basic requirements for composting (Kimura and Shimizu, 1981a,b; Bach et al., 1987; Wu et al., 1990). Composting system technology is required to support production in agricultural ecosystems. However, the main problem is the practical application of these technologies. We begin with an introduction to the composting process (2) and sensor fro systems operation (3), then define with function and mechanism of aeration (4), the results is indicated the results of bin composting (5) and is discussed with the early stage composting by packed bed-type reactor (6) and adiabatic-type reactor (7). Because composting systems are not uniform in degradation and material temperature, information on the degradation of materials within forced aeration composting is very useful for practical operation.

2. THE COMPOSTING PROCESS

Composting is the aerobic (oxygen-requiring) decomposition of organic materials by microorganisms under controlled conditions. During composting, microorganisms consume oxygen (O_2) while feeding on organic matter. Active composting generates considerable heat, large quantities of carbon dioxide (CO_2) and release water vapor into the air. CO_2 and water (vapor) losses can amount to half the weight of the initial waste materials (Fig. 1). Thus, composting reduces both the volume and mass of the raw materials while transforming them into valuable soil conditioner. Factors affecting the composting process are oxygen, aeration, nutrients (carbon:nitrogen (C:N) ratio), moisture content, porosity, structure, texture, particle size, pH and temperature (Table 1).

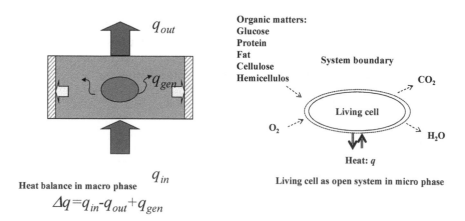

$$\Delta q = q_{in} - q_{out} + q_{gen}$$

Fig. 1. Principles of the composting process.
The carbon, chemical energy, organic matter and water in finished compost is less than that in the raw materials. The volume of the finished compost is 50% or less than that of the raw material.

Table 1. Recommended conditions for rapid composting.

Condition	Reasonable range*	Preferred range
Carbon to nitrogen (C:N) ratio	20:1–40:1	25:1–30:1
Moisture content	40–60%**	50–60%
Oxygen concentration	Greater than 5%	Much greater than 5%
Particle size (diameter in meters)	3.2×10^{-3}–1.3×10^{-2}	Varies**
pH	5.5–9.0	6.5–8.0
Temperature	43–66	54–60

* These recommendations are for rapid composting. Conditions outside these ranges can also yield successful results.
** Depends on the specific materials, particle size, and/or weather conditions.

2.1 Temperature and the Physical Properties of Compost Material

Temperature increase within composting materials is a result of heat balance during composting (Kimura and Shimizu, 2002, Fig. 2a). Temperature is one of the most important variables in the composting process (Schulze, 1962). Composting at temperatures below 20°C has been demonstrated to significantly slow and even stop the composting process. Therefore, temperature can be an indicator of activity in the biological process of composting. In the aerobic decomposition of biomass, the desired products are water, CO_2 and heat byproducts of composting. Mesophilic organisms which function best within the range of 24 to 40°C, initiate the composting process (Fig. 2b). As microbial activity increases soon after the formation of a composting pile, temperatures within piles of sufficient volume and density also increase. Thermophilic microorganisms take

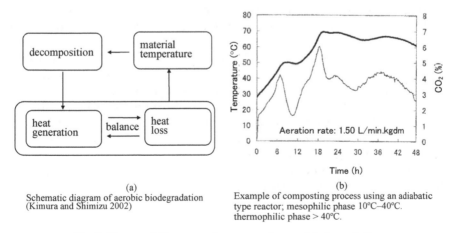

(a)
Schematic diagram of aerobic biodegradation
(Kimura and Shimizu 2002)

(b)
Example of composting process using an adiabatic type reactor; mesophilic phase 10°C–40°C. thermophilic phase > 40°C.

Fig. 2. Thermophilic composting process by aerobic degradation.

over at temperatures above 40°C. The temperature in the compost matrix typically increases rapidly to 54 to 65°C within 24 to 74 h in an adiabatic-type reactor at the laboratory scale (Kimura and Shimizu, 1981a). In thermophilic composting, any soluble sugars in the original mixture are almost immediately used up by bacteria and other microorganisms. Other components such as protein, fat, and cellulose get broken down by heat-tolerant microbes. Nitrogen is readily available when it is in the proteinaceous, peptide, or amino acid forms. Lignins (large polymers that cement cellulose fibers together in wood) are among the slowest compounds to decompose because of their complex structure that is highly resistant to enzyme attack.

Porosity, structure and texture relate to the physical properties of a material such as particle size, shape, and consistency, affect the composting process by their influence on aeration. They can be modified by the selection of raw materials and grinding or mixing. Materials added to adjust these properties are referred to as amendments or bulking agents. For composting applications, an acceptable porosity and structure can be achieved in most of the raw materials, if the moisture content is less than 65% (w/v). However, some situations profit from special attention to porosity, structure, or texture. Composting piles are susceptible to settling, so large particles are necessary. Materials with a strong odor might be mixed with rigid materials to achieve greater than normal porosity in order to promote good air movement.

3. SENSORS FOR SYSTEM OPERATION

In many composting systems, temperature directly activates the aeration devices and is monitored and controlled by sensors for system operation during the initial and final stage of composting (Fig. 3). Aeration is activated or increased when the process temperature surpasses a temperature set point. In other system operation, aeration is determined by a time cycle that is adjusted either manually or automatically according to process temperature. Even with direct temperature feedback control systems, a timer is often required to activate aeration at regular intervals to maintain aerobic conditions when temperature remains below the set point, especially during the initial and final stages of composting (Finstein et al., 1983). Aeration rates and intervals normally vary with the stage of composting (Lenton and Stentiford, 1990). Composting systems can include several temperature zones, each requiring slightly different air flow rates and temperature set points.

	Temperature	O_2	CO_2
Purpose	Heat generation	O_2 consumption	CO_2 generation
Characteristics	High temp.≠ Active reaction	Galvanic cell type sensor ↓ Deteriorated Unsuitable long- time use	Infrared laser sensor ↓ Stable

Fig. 3. Sensors for system operation.

4. FUNCTIONS AND MECHANISMS OF AERATION

The composting method determines how aeration occurs. Aeration is a crucial and inherent component of composting; it provides the O_2 needed for aerobic biochemical processes and removes heat, moisture, CO_2 and other products of decomposition. In the entire composting period, the amount of aeration required for cooling greatly exceeds the amount required for removing moisture or supplying O_2. Thus, the need for aeration is more often determined by temperature rather than by O_2 concentration.

Although there are many variations, aeration generally takes place either passively or by forced air movement. Passive aeration, often called natural aeration, takes place by diffusion and natural air movement. Forced aeration use fans to move air through the mass of composting materials. A third mode of aeration is being developed where nearly pure CO_2 is injected into a closed composting reactor (Rynk and Richard, 2001).

4.1 Passive Aeration

Composting systems that rely on passive aeration normally include periodic agitation or "turning" of the materials. Although turning charges materials with fresh air, however the air introduced is quickly consumed by the composting process (Epstein, 1997; Haug, 1993). The longer lasting effect of turning on aeration might be to rebuild pore spaces in the material, which are crucial for diffusion and convection. However, there is evidence that this effect can be short-lived as well (Michel et al., 1996).

4.2 Forced Aeration

Depending upon the composting systems, forced aeration can be continuous and the rate of aeration can be increased or intermittently turned on and off as needed. Continuous aeration can reduce the required air flow rate. It also reduces the fluctuation in temperature and O_2 levels (Puyuelo et al., 2010) that occur over time. However, continuous aeration can cause gradients within the composting environment leading to excessive drying and permanent cool zones in the area where the air enters (Citterio et al., 1987). This might be a concern if "Process to Further Refuse Pathogens (PFRP)" is required (U.S. EPA, 2016). A process to further reduce pathogens (PFRP) is a treatment process that is able to consistently reduce sewage sludge pathogens (i.e., enteric viruses, viable helminth ova, fecal coliforms, and *Salmonella* spp.) to below detectable level at the time the treated sludge is used or disposed (U.S. EPA, 2016). Forced aeration is typically controlled based on the temperature within the composting materials. Composting experiments under various aeration conditions were performed using an adiabatic-type reactor (Kimura et al., 2007, Table 2).

Table 2. Experimental design setup.

No.	Aeration method	Aeration rate [L/min kgDM]	Length of 1 cylce [hr]	on [min]	off [min]	Total Length of aeration [hr]	Total volume [L/kgDM]
1	Continuous	0.5	-	-	-	48	1140
2		1.5				48	4320
3		3				48	8640
4	Intermittent	0.5	2	80	40	32	960
5				60	60	24	720
6				40	80	16	480
7				20	100	8	240
8		1.5		80	40	32	2880
9				60	60	24	2160
10				40	80	16	16
11				20	100	8	8
12		3		80	40	32	32
13				60	60	24	24
14				40	80	16	16
15				20	100	8	8

5. BIN COMPOSTING

Forced aeration has been used in practical bin composting systems for over two decades. A fan is used to deliver air (oxygen) for aerobic fermentation. Air velocity within the compost material is very low. A diagram of the composting facility in the Nippon Agricultural Research Institute in Tsukuba, Ibaraki Prefecture is shown in Fig. 4. In this facility, four bins (two fermentation bins and storage bins each) are used and the composting materials are moved periodically from one bin to the next in succession. Odorous emissions are sucked by a pump and delivered to a biofiltration facility. The lower limit of aeration is derived from the rate of oxygen consumption for organic decomposition. Peak rates of about 4 to 14 mg O_2/g volatile solid-h were observed in the temperature range of 45 to 65°C (Haug, 1993). Iwabuchi and Kimura (1994) reported that the oxygen uptake rate of dairy cattle manure at a moisture content of 76.7% (w.b.) was 4.8 g/h. kg-(volatile matter: VM). The volatile solid (volatile matter) method estimates organic and ash concentrations. The portion of the sample lost in high-temperature combustion (550°C) estimates volatile matter; the portion remaining after combustion is ash. It has been suggested that aeration rates of 0.20 and 1.33 L/min. kg-VM are suitable for composting mixtures of municipal sewage sludge and garbage, respectively (Lau et al., 1992). Kimura and Shimizu (1989) recommended that air flow for swine manure composting with initial moisture content of 50–60% was 0.3–1.0 L/min. kg-(dry matter: DM). Lau et al. (1992) recommended that aeration at 0.04–0.08 L/min. kg-VM was suitable for swine waste composting.

Kimura and Shimizu (1981b) reported that there were three levels of aeration rates: low, medium and high. At low aeration rates (0–1 L/min. kg-DM), increasing aeration rate increased the maximum temperature, dry

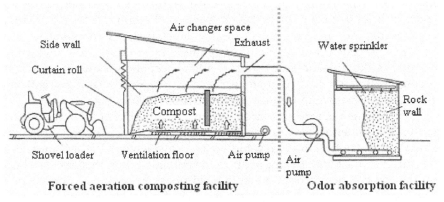

Fig. 4. Bin composting facility.
Nippon Agricultural Research Institute

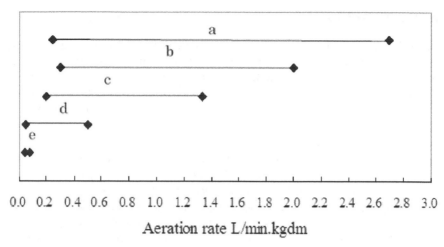

Fig. 5. The range of aeration conditions used in this and previous studies.
Note: a: Kimura et al. (2007), b: Doshu (2003), c: Schulze (1962), d: This work 2005, e: Lau et al. (1992).

matter loss and total weight loss. For medium aeration rates (1–5 L/min. kg-DM), there was higher weight and dry matter loss but the composting temperature was lower than under low aeration rates. With high aeration rates (> 5 L/min. kg-DM), fermentation was slow.

The aeration rates used in previous studies are shown in Fig. 5. Air is supplied not only as source of oxygen for microbial growth in compost but also for other purposes such as control odor released from compost. Turning or mixing during composting is done to minimize the heterogeneity associated with temperature, oxygen and moisture gradients in the system (Vandergheynst and Lei, 2003). Haug (1993) stated that the objective of turning was to reform the compost structure and expose fresh material for microbial colonization. The role of mechanical turning is to increase free air space in order to ensure the highest possible ventilation rate for a particular composting mixture. If a good mix is developed, microorganisms can function efficiently and air will flow through the material more uniformly because of the breakdown of short-circuit air channels. Low nitrogen and organic matter content, high maturity and low viable seed content are associated with turning frequency (Anonymous, 2005b, Fig. 6).

The problem in forced aeration systems is uniformity of decomposition within the material. Since air generally flows from the bottom (plenum chamber), the bottom layer has high oxygen availability. Bottom layer also has direct contact with the input air from ventilation floor that released heat by the aerobic biodegradation from solid phase in compost. Composting also has the objective of killing pathogens. The United States Environmental Protection Agency suggests that the minimum operating temperature must

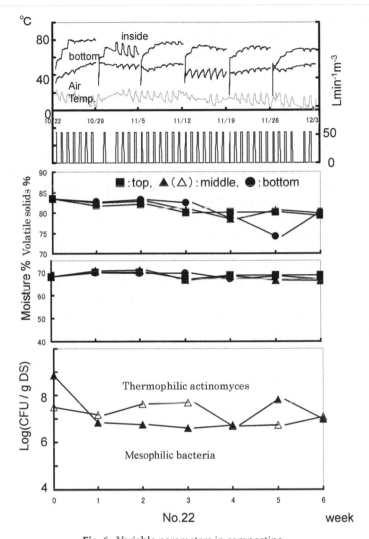

Fig. 6. Variable parameters in composting.

be maintained at 53°C for 5 days, 55°C for 2.6 days and 70°C for 30 min (Lau et al., 1992). The minimum temperature for composting is 50°C.

6. PACKED BED-TYPE REACTOR

The early stage of composting is important in determining the success of the process. It is desirable for the temperature to increase substantially in the early stage of composting since temperature is important for microbial activity. The objective of this study is to determine the effect of aeration on the decomposition process and odor emission in cattle manure composting during the early stages of composting (0–120 h). The characteristics of the compost material (including change in moisture content, total weight loss, and odor generation) during composting were studied.

6.1 Compost Reactor Setup

The compost reactor setup is shown in Fig. 7. With a total capacity of 18.84 L, the compost reactor was fabricated using Ø 20 cm of polyvinyl chloride (PVC) pipe of 15 cm height per layer; with four layers stacked on top of each other. The bottom layer was the plenum chamber. Each layer had wire mesh (1 × 1 mm) at the bottom which facilitated easy sampling within the layers without mixing with other layers. Different turning patterns could be experimented within the compost reactor. The reactor could receive forced aeration through a Ø 5 mm opening from an air pump (DAP 30, ULVAC KIKO, Yokohama, Japan) with a capacity of 30 L/min. Air flow meters (Model RK1400 series, KOFLOC, Kyoto, Japan) with capacities of 0.5 and 2.0 L/min were used. The outer surface and bottom of the compost reactor were insulated with wool and fiberglass of 10 cm thickness to reduce heat loss. Approximately 8 kg of mixed cattle manure-sawdust were placed in the compost reactor. Temperatures were measured with T type thermocouples which were inserted via small holes (4 mm) in the PVC pipe in each layer. Temperature data were recorded with a data

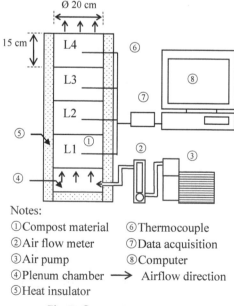

Notes:
①Compost material   ⑥Thermocouple
②Air flow meter   ⑦Data acquisition
③Air pump   ⑧Computer
④Plenum chamber   ⟶ Airflow direction
⑤Heat insulator

Fig. 7. Compost reactor setup.

recorder (NR 1000, Keyence, Osaka, Japan) at 30 min intervals. A personal computer was connected to the data loggers to record and store data.

The average temperature of the compost material in the each four layers (L1, L2, L3, L4) at different aeration rates is shown in Fig. 8. The maximum average temperatures at aeration rates of 0.025, 0.050, 0.100 and 0.150 L/min. kg-DM were 37.4, 49.3, 63.5 and 49.1°C, respectively. The average temperature was highest at 0.100 L/min. kg-DM aeration rate. Aeration rate of 0.025 L/min. kg-DM led to poor composting in terms of temperature. Lau et al. (1992) reported that temperatures during passive aeration could reach up to 65°C. In this study, the reactor was not suitable for passive aeration.

Odor is a product of the decomposition of organic materials. The type of gas produced reflects the condition of the composting process. Ammonia (NH_3) is a product of aerobic fermentation, while hydrogen sulfide (H_2S) is a product of anaerobic fermentation. The highest concentration of ammonia occurred after 48 h (Fig. 9). The third layer had the highest concentration of NH_3 at 7000 ppm, followed by the second, first and top layers. This means that microbial activity was the highest in the third layer. When the aeration rate was low, NH_3 production was also low. At 0.05 L/min. kg-DM aeration rate, the first layer had the highest NH_3 (120 ppm) after 72 h, followed by the second, third and fourth layers. At the lowest aeration rate of 0.025 L/min. kg-DM, the concentration of NH_3 was 1 ppm. It is clear that the supply of sufficient air (up to 0.100 L/min. kg-DM) resulted in high ammonia generation.

About 8 kg of mixed cattle manure-sawdust was placed in the compost reactor. The mixed cattle manure-sawdust was composted without adjusting the pH of the initial mixture. Each layer had a capacity of 2 kg.

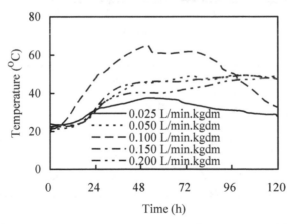

Fig. 8. Effect of aeration rates on average temperature of compost material during 120 h of composting.

Three aeration rates (0.05, 0.15 and 0.50 L/min. kg-DM) were applied for each run. Samples from each four layers were collected every 120 h before turning to measure the moisture and ash contents. Data was collected from March to July 2005.

Three types of turning pattern were used in the study: without turning (control) (Run A), full turning (Run B) and turning of each layer

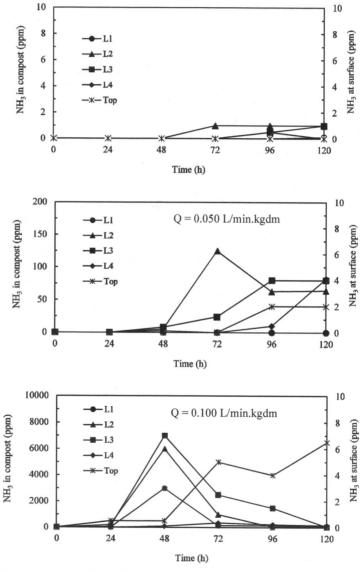

Fig. 9. Concentration of NH_3 in different layers during 120 h of composting.

and change in position of layers (Run C). In the third run, there was a special turning regime that involved changing the position of layers during different stages of composting (Karyadi et al., 2007, Fig. 10). The experiment with change in position of layers was possible because the reactor had wire mesh at the bottom of each layer which made it possible to prevent the material in one layer from mixing with that of another. In run

A. No turning

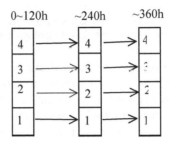

B. Full turning

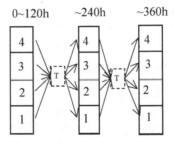

C. Turning and position change

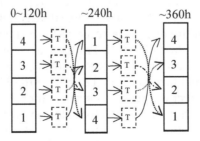

Fig. 10. Schematic diagram of the turning pattern for runs A, B, and C (T: Turning; 1, 2, 3 and 4: layer number).

Table 3. Experimental design setup for cattle manure composting with two factors (aeration rate and turning pattern).

Turning method	Aeration rate (L/min. kgdm)		
	0.05	0.15	0.5
A	0.05A	0.15A	0.50A
B	0.05B	0.15B	0.50B
C	0.05C	0.15C	0.50C

Aeration rates 0.05 to 0.50 L/min. kgdm (to achieve high temperature in composting process) Turning pattern: no turning (A), full turning (B) and turning with position change (C).

A, the sample was left without disturbance, except for sample collection. In run B, after 5 days the samples were collected and the compost material from each layer placed in a plastic bucket; the material was then turned with a scoop and returned to the same layer. In run C, just after sample collection, the compost material was taken out from each layer and then returned to the same vessel after turning; but the position of the layers was reversed from the previous period. Since the direction of air supply was from the bottom, the change in position of layers might provide a suitable alternative turning method. The experimental design setup for these experiments is shown in Table 3.

6.2 Temperature Distribution

Temperature distribution in the compost materials during 360 h of composting for all runs is shown in Fig. 11. Temperature is the result of the decomposition of organic matter. When the heat balance of the system is positive, the temperature of the material increases. When compared with a small adiabatic reactor system, this reactor has more uncertainties. Factors affecting the results are: the freshness of the manure, type of sawdust, initial moisture content and the degree of mixing of the sample. Since it is difficult to use the composting same raw material for each runs on a large scale, the raw material was prepared for each runs. This meant that the freshness of the raw material was not uniform. The initial moisture content of the manure also varied. The minimum amount of sawdust was added to obtain 65% (w.b.) moisture content in the final compost mixture.

Maximum temperatures were recorded during the first period of the the compost turned every 120 h over the 360 h period. The turning pattern did not affect the results, because after the first turning, for both types of turning patterns, temperature decreased below that obtained in the first period. The maximum temperatures at aeration rates of 0.05, 0.15 and 0.50 L/min. kg-DM were 64.0, 73.2 and 70.8°C, respectively. In each run, the time of peak temperature was different for each layer. In run A, with .05 aeration

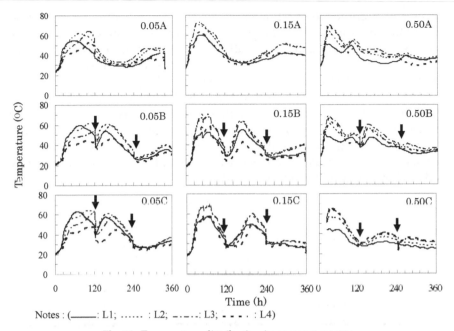

Fig. 11. Temperature distribution in compost reactor.

(.05A), the maximum temperature 64.3°C occurred in the third layer at 96 h and increased again at 216 h. This indicated that the decomposition process increased again after the first peak from the beginning of the composting process. Run 0.05B showed a different temperature distribution as the temperature increased after turning which indicated that the decomposition rate had increased again due to the breakdown of the structure of the compost material during turning. In runs 0.05B and 0.05C, the temperature increased in the first and second layers faster than in the other layers because of the availability of oxygen supply.

For aeration rate of 0.15 L/min. kg-DM, maximum temperature was reached after 37 h of composting. In run 0.15A, the temperature increased slightly after 192 h; in run 0.15B the maximum temperature was reached after 67 h while in run 0.15C, the second highest temperature was obtained after a change in position after turning at 192 h. The temperature in the first layers increased faster than in the other layers, even though the maximum temperature obtained was lower. The differences in time at which maximum temperatures were reached for runs 0.15A and 0.15B was caused by the differences in freshness of the raw material.

For aeration rate of 0.50 L/min. kg-DM, the maximum temperature of 70.8°C after 24.5 h of composting indicate that aeration had a substantial effect on heat production. In run 0.50A, temperature increased again at 72 h. In the second and third periods of turning, the temperature decreased in all layers except for the first layer. In run 0.50C, temperature distribution was different from that in the previous treatment. Temperature slightly increased and then decreased, until the next turning period. Generally, with full turning, the material is more uniform after turning, thus, temperature distribution was similar to that in the previous period.

The mass balance during composting is shown in Fig. 12. Calculations were made with the assumption that ash content is an inert material during the composting process and only organic matter and level of water changed. Turning affected dry matter reduction, however, total mass reduction was most likely affected by aeration. Fermentation and drying reduced the mass of wet and dry matter.

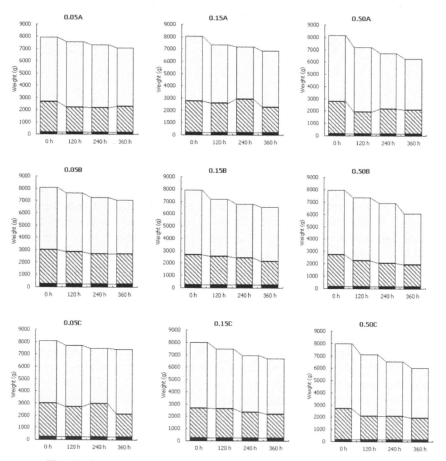

Fig. 12. Composition of water, organic matter and ash during composting.
(▭: water, ▨: organic matter, ▬: ash)

7. ADIABATIC-TYPE REACTOR

The problem in composting with a laboratory-scale reactor is heat loss to the surroundings (Bach et al., 1987). To overcome this problem, some researchers have made small reactors with an adiabatic system (Kimura and Shimizu, 1981a). It is easy to obtain precise data to observe the composting process while using this kind of reactor. Many studies have used this kind of reactor to explain the mechanisms of composting. In laboratory-scale compost reactors, effects of the surroundings must be considered. Bach et al. (1987) explained water migration within compost reactors. Shimizu et al. (1989) developed a model of drying and fermentation in a compost reactor.

In small compost reactors, processes within the material are more uniform than in larger reactors. Kimura and Shimizu (1981a) developed an adiabatic-type reactor. To conduct satisfactory experiments with a compact device, an adiabatic system was adapted for a compact composting container, where the surrounding temperature of the container was automatically maintained at the same temperature as that of the internal material. As a result, the material temperature increased as high as that obtained in a large-scale practical composting system. Various aeration treatments were tested to achieve low cost and verify safety aspects of composting.

7.1 Compost Reactor Setup

An adiabatic-type compost reactor was used in these experiments (Fig. 13). The compost reactor consisted of: (1) temperature controller, (2) reactor, (3) aeration device, (4) ammonia trap, and (5) digital recorder.

1) *Temperature controller*

   Temperatures inside and outside of the reactor were measured using the platinum resistance thermometer sensor (SCYS, CHINO, Tokyo, Japan). The difference in temperature between the surrounding reactor and the compost material was maintained at < 1.5°C with a heater.

2) *Reactor*

   The reactor had a volume of 700 mL and was fabricated from glass material. 250 g of raw material (manure) was placed in the reactor. The reactor was insulated with Styrofoam.

3) *Aeration device*

   A timer was connected to the air pump to regulate the aeration system (on/off mode). The air was filtered through a micro filter (FLHN type) while a flow meter was used to adjust the aeration rate.

4) *Ammonia trap*

   The air flow released ammonia and water vapor. For measuring the concentration of CO_2, ammonia was trapped using 0.1N of H_2SO_4.

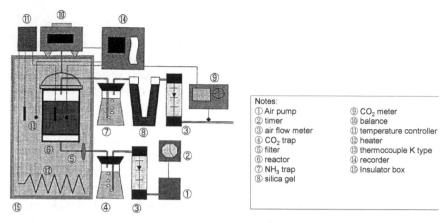

Fig. 13. Compost reactor setup.

Water vapor was reduced by delivering air to a silica gel tube before the CO_2 meter.

5) *Measurement devices*

A data recorder (Process VII, CHINO), thermocouple type K, digital balance (PD1-2400 W, Chou Balance), and portable CO_2 meter (CGP-1, TOA) were used in these experiments. Data was recorded in time intervals of 10 min.

7.2 Data Analysis

The energy in composting was classified into enthalpy, ΔH_g and energy for catabolism, ΔH_{ca}.

$$\Delta Q = \Delta H_g + \Delta H_{ca} \tag{1}$$
$$\Delta Q = \Delta H_2 + \Delta H_1 + \Delta H_3 \tag{2}$$

ΔH_g was neglected because it is very small compared to ΔH_{ca}. ΔH_1, ΔH_2, and ΔH_3 as described in equations (3), (4) and (5). The reactions associated with aerobic metabolism are shown below: Reactions (3) and (5) represent the degradation of organic matter while reaction (4) represents the synthesis of organic matter in microbes.

$$C_xH_yO_z + (x + y/4 - z/2)O_2 \rightarrow xCO_2 + y/2H_2O, \quad \Delta H_1 \tag{3}$$

$$n(C_xH_yO_z) + nNH_3 + n(x + y/4 - z/2 - 5)O_2 \rightarrow$$
$$(C_5H_7NO_2) + n(x{-}5)CO_2 + n/2(y{-}4)H_2O, \quad \Delta H_2 \tag{4}$$

$$C_5H_7NO_2 + 5nO_2 \rightarrow 5nCO_2 + 2nH_2O + nNH_3, \quad \Delta H_3 \tag{5}$$

ΔH_2 is the energy produced during synthesis in the microbial body. Because ΔH_3 is $1/160$–$1/120$ of ΔH_1, ΔH_3 which is also neglected (Kimura and Shimizu, 1989).

$$\Delta W_m = \Delta W_d + \Delta W_w \qquad (6)$$

where ΔW_m = weight loss, ΔW_d = dry matter loss and ΔW_w = water loss

$$\Delta W_d + \Delta O_2 = \Delta CO_2 + \Delta H_2O \qquad (7)$$

where ΔO_2 = oxygen, ΔCO_2 = carbon dioxide and ΔH_2O = water production

$$\Delta W_{tw} = \Delta W_w + \eta \times \Delta W_d \qquad (8)$$

where ΔW_{tw} = total water loss and $\eta = 0.56$–0.60

$$\Delta W_w = \Delta W_{tw} - \eta \times \Delta W_d \qquad (9)$$

For analysis, weight loss, organic matter loss, and moisture loss are calculated as follows:

1) *Weight loss*

$$\text{Weight reduction (\%)} = \Delta W_m / W_0 \times 100 \qquad (10)$$

where W_0 = the initial weight

2) *Organic matter loss*

From the CO_2 concentration data, the weight of dry matter loss ΔW_d (g) could be calculated as follows:

$$\text{Organic matter loss (\%)} = \Delta W_d / W_0 \times (100 - m_0) \times 100 \qquad (11)$$

where m_0 is the initial moisture content in % (w.b.).

3) *Moisture loss*

$$\text{Moisture loss (\%)} = \Delta W_m / (W_0 \, x \times m_0 / 100) \times 100 \qquad (12)$$

4) *Calculation for intermittent aeration and energy saving*

Total aeration (L/min. kg-DM) = Aeration rate (L/min. kg-DM) × length of aeration × 60 $\qquad (13)$

7.3 Temperature and CO_2

The temperature profile and CO_2 concentration during continuous aeration is shown in Fig. 14. Temperature at an aeration rate of 0.50 L/min. kg-DM reached more than 60°C (Kimura and Shimizu, 1981a). Two peaks of CO_2

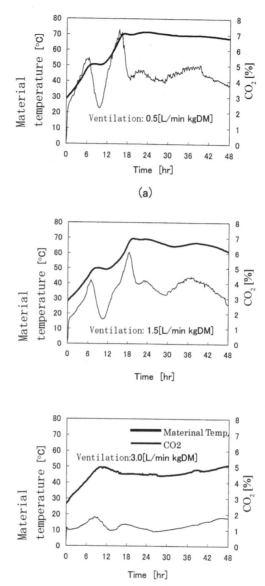

Fig. 14. Changes in material temperature and CO_2 concentration for composting with continuous ventilation.

concentration indicated that two stages of composting (mesophilic and thermophilic) occurred in the composting process. The phase changes of these stages were clear at aeration rates of 0.50 to 1.50 L/min. kg-DM. The thermophilic stage was associated with high microbial activity and refuse pathogen aspect. The maximum temperatures and times required to reach the maximum temperature at aeration rates of 0.50, 1.50 and 3.00 L/min. kg-DM were 71.4, 70.0 and 49.1°C and 24.0, 19.6 and 10.9 h, respectively. Aeration rates of 0.50 and 1.50 L/min. kg-DM resulted in high temperature composting. The time required to reach the maximum temperature could be reduced by increasing the aeration rate from 0.50 to 1.50 L/min. kg-DM. Excess aeration occurred at an aeration rate of 3.00 L/min. kg-DM resulting in lower temperatures at other aeration rates.

This chapter focused on process optimization of composting system. Especially, forced aeration has been used in practical bin composting systems for over two decades. But ideal operational conditions of composting are still being researched. Thus, the works and challenges for the future are (1) to determine the appropriate strategy for cattle manure composting in packed bed reactor with forced aeration and (2) to examine the aeration conditions at early stage of composting influence decomposition of organic matter for compost and to combine sensor technique during the composting reaction with knowledge of aeration rate, aeration method (continuous, intermittent) and turning method to determine the optimal conditions of composting systems.

8. SUMMARY

Aerobic fermentation was chosen to solve the cattle manure waste problem because it is a simple low energy solution. The combination of different aeration rates and turning methods yielded different results. Studies using a packed-bed type reactor confirmed that degradation in forced aeration composting can be increased with a combination of full turning or turning with a position change. The huge amount of solid waste from animal husbandry results in odorous emissions, thus composting operations must be automated. However, composting systems are complicated by lack of uniformity in degradation and material temperature. Future research should focus on further development of sensors for composting system operations and control systems also in addition to basic research on composting processes to support the development of "smart agriculture". Development of composting system technologies will contribute to the "sustainable development of agricultural production".

REFERENCES

Anonymous. 2005b. Improving and maintaining compost quality (Compost Factsheet #3). Cornell Waste Management Institute. http://cwmi.css.cornell.edu (accessed in January 2017).
Bach, P. D., Nakasaki, K., Shoda, M., and Kubota, H. 1987. Thermal balance in composting operations. J. Ferment. Technol. 65(2): pp. 199–209.

Citterio, B., Civilini, M., Rutili, A., and de Bertoldi, M. 1987. Control of a composting process in bioreactor by monitoring chemical and microbial parameters. pp. 633–642. *In:* de Bertoldi, M., Ferranti, M. P., L'Hermite, P., and Zucconi, F. (eds.). Compost: Production, Quality and Use. Elsevier Applied Science, London, United Kingdom.

Epstein, E. 1997. The Science of Composting. Technomic Publishing, Basel, Switzerland.

Finstein, M. S., Miller, F. C., Strom, P. F., MacGregor, S. T., and Psarianos, K. M. 1983. Composting ecosystem management for waste treatment. Bio/Technol. 1: 347–353.

Haug, R. T. 1993. The Practical Handbook of Compost Engineering. Lewis Publisher.

Iwabuchi, K., and Kimura, T. 1994. Aerobic biodegradation of dairy cattle feces (Part 1) (in Japanese). J. of JSAM. 56(2): pp. 67–74.

Karyadi, J. N. W., Harano, M., Shimizu, N., Takigawa, T., and Kimura, T. 2007. Degradation of organic matter in cattle manure composting by aeration and turning in a packed-bed reactor. J. of JSAM. 69(4): pp. 71–78.

Kimura, T., and Shimizu, H. 1981a. Basic studies on composting of animal waste (1) (in Japanese). J. of JSAM. 43(2): pp. 221–227.

Kimura, T., and Shimizu, H. 1981b. Basic studies on composting of animal waste (2) (in Japanese). J. of JSAM. 43(3): pp. 475–480.

Kimura, T., and Shimizu, H. 1989. Basic studies on composting of animal waste (3) (in Japanese). J. of JSAM. 51(1): pp. 63–70.

Kimura, T., and Shimizu, N. 2002. Performance evaluation of the composting facility for producing high quality compost by means of the natural energy (in Japanese). Annual Report for BRAIN. pp. 1–16.

Kimura, T., Shimizu, N., Kayadi, W. N. J., and Sato, K. 2007. Effect of intermittent ventilation on composting of cattle manure with a small-scale reactor (in Japanese). J. JSAM. 69(5): pp. 78–87.

Lau, A. K., Lo, K. V., Liao, P. H., and Yu, J. C. 1992. Aeration experiment for swine waste composting. Bioresour. Technol. 41: pp. 145–152.

Lenton, T. G., and Stentiford, E. I. 1990. Control of aeration in static pile composting. Waste Manage. Res. 8: pp. 299–306.

Michel, F. C., Forney, L. J., Jr., Huang, J. -F., Drew, S., Czuprenski, M., Lindeberg, J. D., and Reddy, C. A. 1996. Effects of turning frequency, leaves to grass ratio, and windrow vs. pile configuration on the composting of yard trimmings. Compost Sci. Util. 4(1): pp. 26–43.

Puyuelo, B., Gea, T., and Sánchez, A. 2010. A new control strategy for the composting process based on the oxygen uptake ratio. Chem. Eng. J. 165: pp. 161–169.

Rynk, R., and Richard, T. L. 2001. Commercial Compost Production Systems. pp. 51–120. *In:* Stoffella, P. J., and Kahn, B. A. (eds.). Compost Utilization in Horticultural Cropping Systems. CRC LLC, New York.

Schulze, K. L. 1962. Continuous thermophilic composting. Appl. Microbiol. 10(2): pp. 108–122.

Shimizu, H., Wu, X., Sato, K., Nishiyama, Y., and Kimura, T. 1989. Heat and mass transfer in the aerobic fermentation and drying process of organic material in packed bed. J. Soc. Agricul. Struc. 20(2): pp. 169–176.

United States EPA. 2016. Basic Information: Pathogen equivalency committee. https://www.epa.gov/biosolids/basic-information-pathogen-equivalency-committee. (accessed in January 2017).

Vandergheynst, J. S., and Lei, F. 2003. Microbial community structure dynamics during aerated and mixed composting. Trans. of the ASAE. 46(2): pp. 577–584.

Wu, X., Shimizu, H., Nishiyama, Y., and Kimura, T. 1990. Effect of aeration rate on the aerobic fermentation and drying process of organic waste material in packed bed. J. Soc. Agricul. Struc. 20(3): pp. 230–236.

2

Overview of Mechatronic Design for a Weed-Management Robotic System[#]

T. Perez,[1,2,*] *O. Bawden,*[1] *J. Kulk,*[1] *R. Russell,*[1] *C. McCool,*[1] *A. English*[1] *and F. Dayoub*[1]

1. INTRODUCTION

This chapter discusses key aspects of a design of a robotic platform for the management of crops in agriculture. In particular, the system considered seeks to address the increasing threat of weed species resistant to herbicide. Many crop production systems in countries such as Australia, have moved onto non-tillage practices in order to reduce loss of soil moisture and soil nutrients to the atmosphere. These crop productions rely mostly on chemical agents as a means of weed management. Such a practice has contributed to the development of weed species for which chemical agents have lost they effectiveness—this is known as weed resistance.

The control of weeds is an important aspect of farming. A weed is any plant species in the paddock that is not the current crop. Weeds use soil moisture and nutrients and therefore compete with the crop. Some

[1] Electrical Engineering & Computer Science, Queensland University of Technology, Brisbane, Australia.
[2] Institute for Future Environments, Queensland University of Technology, Brisbane, Australia.
* Corresponding author: tristan.perez@qut.edu.au
[#] This research was co-sponsored by QUT (Faculty of Science and Engineering and The Institute for Future Environments) and The Queensland Government through its Department of Agriculture and Fisheries (DAF) under the Strategic Investment in Farm Robotics Program. Their sponsorship is dully acknowledged.

weed species can even release chemicals that inhibit the crop growth. Furthermore, the presence of weeds at harvesting can affect the quality of the harvesting and that of the harvested crop. Historically, weed management has been conducted using mechanical means of removing the weeds.

A contribution of the development of machinery through and since the industrial revolution combined with the development of synthetic fertilisers and improved crop breeding programs during the green revolution (1960s to 1980s) have resulted in an increase in the size of fields or broadacre farming and the use of chemicals as the prevalent weed control agent. The development of herbicide resistant weed species is of great concern to farmers as it poses a threat to future crop production (International survey of herbicide resistant weeds, www.weedscience.org).

The use of robotic vehicles that can autonomously manage the weeds can offer a potential solution to this problem as they enable the use of alternative weed-control methods. Over the past twenty years there has been a small pocket of activity in research related to alternative methods for weed destruction; whether mechanical, thermal or radiation based. We could argue that this research has not evolved into widespread use due to both the effectiveness and the economics of chemical agents (Upadhyaya and Blackshaw, 2007). Nevertheless, weed resistance is changing this. Robotic weeders, can operate in groups and thus reduce the speed of each machine without increasing significantly the time required for the weeding operation. This reduction in speed creates an opportunity for alternative and energy efficient methods. Robotic weeders have another two key beneficial side effects.

The first is the reduction in robot size and weight, which leads to reduction in soil compaction that adversely affects the crop root development and subsequently its yield. The second, is that whilst traversing the paddock robotic weeders can carry sensors to collect data that can be used in other aspects of crop management; for example, nutrition, water stress, pest and diseases.

This chapter discusses aspects of mechatronic design of the robotic system for weed management depicted in Fig. 1. Such a system consists of a robotic platform, named AgBot II, which can autonomously navigate in the paddock, a vision-and-weeding implement sub-system to manage weeds, and a replenishment pod sub-system for energy and, in some cases, other inputs. This chapter concentrates only on aspects of design that involves a mechatronics approach—this is elaborated in the next section.

Fig. 1. Robotic system for weed and crop management. AgBotII robotic platform with underhanging weeding implement and replenishment pod.

2. A MECHATRONIC DESIGN APPROACH

Mechatronic design deals with the complete design of a mechatronic system rather than single components. Mechatronic systems have a mechanical (physical) component whose desired motion behaviour is controlled through the use of force actuators commanded by a computer control system that processes information from data generated by sensors. This is illustrated in Fig. 2.

The main characteristic of a mechatronic design approach is that all components of the system (mechanical, actuators, sensors, computers, and control) are considered *ab initio* as part of the design. As such, this can address complex interactions among the decisions made in the sub-design of each of the system components, and lead to superior performance, economy, and safety. This requires a multi-disciplinary knowledge since the decisions made in each component of the system often affect that of the other components and together contribute as factors, of affecting overall performance, economy and safety. For example, decisions made about mechanical design affect the dynamical characteristics of the mechanical component, and hence the potential complexity of the motion controller structure. On the other hand, the desired closed-loop behaviour and the dynamics of the mechanical components determine the energy

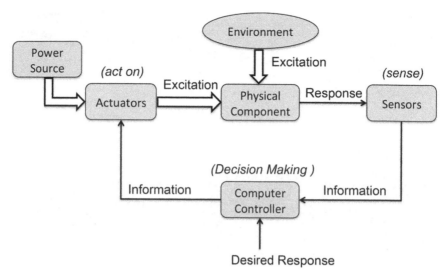

Fig. 2. Structure of a mechatronic system.

requirements and thus the necessary actuation, which can affect both the choice of materials and the need to distribute mass so as to avoid unwanted resonance frequencies of mechanical vibration and material fatigue. The dynamic response, the structure of the control system, and the required sensing modalities put a constraint on the computational speed at which information needs to be processed in order to compute and implement the control action and at the same time conduct condition monitoring of the system for adequate safety shutdown or fault-tolerance.

In the case of the robotic system shown in Fig. 1, there are two key mechatronic systems: the robot platform (vehicle) and weeding implement, the design of which are not independent of each other and have their own associated challenges. For example, the desired motion behaviour of the robotic platform is characterised by its speed and heading rate so it can follow desired paths as determined by an operation planning system that takes into account the geometry of the paddock, the number of robots in the paddock, and the orientation of the crop rows. The design of the platform wheel and driving-wheel configuration influence the number of actuators and complexity of the motion control architecture and control algorithms. The mass and mass distribution determine the dynamic response characteristics and the power requirements. The speed of the operation should be considered in conjunction with the design of the weeding implement's capabilities in order to achieve high weeding efficiency and satisfy constraints on the computational time required by a computer vision system to process data from multiple images to detect and classify the weed species and then decide on the control method.

The rest of the chapter elaborates the mechatronic design process and its components and highlights the interaction among the decisions made in each component.

3. SYSTEM SPECIFICATIONS

From the inception and throughout the project, our design team has been interacting with farmers in order to collect information and execute a design that properly addresses the crucial problems. The insights from potential users were used to establish the functional and operational requirements which must balance the complex demands of the system. The incorporation of a user-centred design methodology (Gulliksen et al., 2003) helped uncover key insights during the development of the platform, which we call AgBotII.

3.1 A Farmer's Perspective

Redhead et al. (2015) reports our initial work in which growers (farmers) and agronomists participated in contextual interviews and observational studies at farm locations in the Darling Downs and Emerald regions in Queensland, Australia. The key findings are summarised below:

- AgBots are most suited for precision work that requires accuracy and is difficult to achieve using large machinery, such as 48-metre boom sprayers;

- Growers are interested and competent in thinking through problems related to building mechanical components, and participatory engagement of farmers in the design, testing, and evolution of AgBot prototypes would be beneficial to both the farming community and this research;

- The built of the mechanical components of the system should remain open for ongoing maintenance and adaptability;

- Varying levels of access to the interface system are necessary, with a simple level of control available for non-skilled labour, and more complex levels of administration by farm managers;

- Rural communication infrastructure can not be assumed to the adequate for reliable communication with the robots, and should be addressed as part of the design of robots;

- The data collected by AgBots should be relevant to the scale of the operation and should be stored in manageable data package sizes;

- Growers welcome an open source community model for the software development of AgBots and this should be set up early and in a way to encourage participation from farmers;

- The number of AgBots monitored per operator needs to be manageable in terms of the workload of the operator;

- Remote views of AgBots should give adequate and easily interpreted visual information about the state of the machine and nature of failure modes.

3.2 Functional and Operational Specifications

From the research and insights gained from field studies with farmers, a list of Functional Requirements (*FR*) were identified (Bawden, 2015):

FR_1—The robot must be a multi-role, lightweight platform suited for autonomous operations related to weed management, fertiliser application, and crop scouting;

FR_2—The robot must be able to conduct weed destruction operations in-fallow as well as in-crop, without damaging the crop;

FR_3—The robot must be able to identify (detect and classify) weeds in order to select and apply the most appropriate weed treatment, which includes the integration of chemical and non-chemical destruction methods;

FR_4—The robot must be able to conduct autonomous operations with appropriate levels of safety;

FR_5—The robot must be able to conduct replenishment operations of energy and agricultural inputs with appropriate levels of safety;

FR_6—The robot must be able to self-diagnose failure modes, and either reconsider operation or shut-down safely;

FR_7—The robot must be able to provide adequate levels of communication and human-robot interactions.

Also a list of Operational Requirements (*OR*) were identified (Bawden, 2015):

OR_1—The robots must be able to be transported safely using standard size road vehicles (trailer/truck);

OR_2—The robots must be able to shut-down operation and stop motion if a human comes in close proximity;

OR_3—The robots must be able to shut-down spraying operations if adverse weather conditions eventuate;

OR_4—The robot must be able to sustain operations under appropriate environmental conditions (temperature, humidity, UV) and operational conditions (terrain gradient, soil conditions).

3.3 Technical Specifications

Based on the functional and operational requirements and input, we conducted a detailed trade-off study and determined the main *technical specifications* shown in Table 1.

Vehicle mass (without payload) (TS_1): Consulted farmers suggested that all-terrain-vehicles (ATVs) used for farming should result in minimal soil disturbance when driven over fields in varying conditions. ATV's range in sizes and mass from 200 to 600 kg; therefore, based on this and the required strength, we estimate the target mass to be 500 kg for the vehicle.

Payload mass (TS_2): To determine the mass of the payload, we considered weed management with current spot-spraying technology. Based on a 3 m width, an operational speed of 5 km/h, and spray rate of 15 L/ha a 200 L tank would require refilling every 10 hours. Given the inclusion of the mechanical weeding array, we expect to use less than 15 L/ha, however, the payload must also account for the development of future implements, so a target weight for the payload of 200 kg was adopted.

Operational speed (TS_3): The operational speed of 5 km/h was selected, keeping in mind safety, area under coverage (based on an operational cost model) and time required for the computer vision system to process

Table 1. AgBotII Main Technical Specifications.

	Specification	Magnitude (target)	Unit
TS_1	Vehicle mass	500	kg
TS_2	Payload mass	200	kg
TS_3	Operational speed	5	km/hr
TS_4	Maximum speed	10	km/hr
TS_5	Number of wheels	4	-
TS_6	Drive wheels	2	-
TS_7	Steering wheels	2	-
TS_8	Wheel width	0.3	m
TS_9	Width (wheel centre to wheel centre)	3	m
TS_{10}	Length total	2.5	m
TS_{11}	Implement section clearance	0.75	m
TS_{12}	Operating gradient pitch	15	%
TS_{13}	Operating gradient roll	10	%
TS_{14}	Handle emergency brake	-	-

data for weed detection and classification. The average walking speed of a human is 5 km/h, so humans could easily overtake the robots. The integration of a multi-mode weed management system, designed based on a requirement that the robot does not stop, was also taken into account for the specification of the operational speed.

Maximum speed (TS_4): The maximum speed is selected for traveling to recharging/replenishment stations and to move from paddock to paddock. Given the operational speed, the maximum target speed has been determined based on potential use of gearboxes and hydraulic drive trains and the range of speeds of electrical motors, which are optimised for maximum efficiency at the selected operational speed.

Vehicle configuration (TS_{5-7}): Analysis of vehicle configurations, shown in Fig. 3 took into account manoeuvrability, stability, locomotion type (tracks vs. wheels), the number of drive and steering motors, and motion control design. A four-wheel configuration, capable of bi-directional driving through the use of differential steering and caster wheels (configuration 9 in Fig. 3) was selected for further development. This configuration offers an appropriate balance between driving performance, stability, payload capacity and complexity. In addition, this configuration simplifies the motion controller since the rotation about the vertical axis (vehicle heading) can be decoupled for control design purposes.

Vehicle dimensions (TS_{8-11}): A standard row width of 0.5 m would be appropriate for a large portion of broadacre applications. With allowance

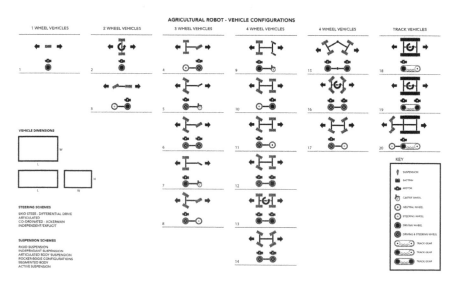

Fig. 3. Twenty vehicle configurations tracked and wheeled variants.

for overhanging leaves and drift in steering, a working width of 300 mm was considered safe for the wheel unit. Australian roads (NTC, 2015) specifies that the width of a vehicle to be moved must not be over 2.5 m. Hence, to move the AgBotII on a typical flatbed trailer the length (or width, depending upon the orientation of the AgBotII to the truck/trailer) needs to be less than 2.5 m. Based on current control traffic farming (CTF) practices a vehicle width (wheel centre to centre) of 3 m was chosen. This enables the AgBotII to take advantage of CTF layout already in place on many broadacre farms. Being wider than the width permissible for carriage on a public road means that the vehicle would need to be loaded perpendicular to the truck or trailer used for carrying the AgBotII. Broadacre crop heights vary according to region, crop variety, moisture availability, soil nutrients and weed competition. Based on average crop-heights of wheat, barley, sorghum, oats, cotton and chickpea, we determine that the vehicle clearance should be at least 0.75 m, with the option for small adjustments based on suspension settings.

Operation (TS_{12-14}): Broadacre farming is generally undertaken on relatively flat terrain. Operating gradients of 0–3% are very common. Paddocks with 5–10% gradients are less common because at this gradient the land acts more energy intensive to cultivate. Based on farm research, an operating gradient of 15% was estimated as the worst case the agricultural robot would see in field conditions. In terms of side constrain forces of wheels, a 10% gradient is considered to be worst case for roll angle operation. It is also a technical requirement that the vehicle must not tip over during emergency braking in gradients of ±15%.

4. AGBOT II DESIGN

As part of the design of AgBot II, we considered different subsystems as indicated in Fig. 4. Different options were analysed for each subsystem taking into consideration their impact on the complete design. In this section, we summarise the design process and some of the features of the AgBot II platform.

4.1 Chassis Design

AgBot II is based on two modular side assembly units joined by the implement unit—see Fig. 4. The side units are symmetrical (mirrored). These units house the battery and power systems and limed to the rear caster wheels as well as the swing arms that connects to the drive unit assembly.

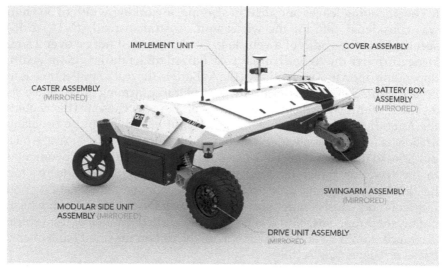

Fig. 4. AgBotII platform rendering showing the major assemblies.

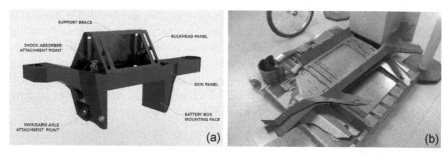

Fig. 5(a) Detail of the side unit assembly for the AgBotII. (b) Flat-pack components for the side unit prior to assembly by MIG welding.

Based upon our interactions with farmers, a novel construction method was developed which takes advantage of the considerable manufacturing infrastructure that already exists on most large farms in Australia. The design incorporates CNC laser cutting, pressing and machining, at low volumes and low cost, to rapidly produce prototypes or kits that could be shipped to farms and assembled on-site by the farmers themselves. Using only a MIG welder, pneumatic rivet gun and simple hand tools, the mechanical portion of the robot chassis itself can be assembled in less than 8 hours by two people.

The implement unit shown in Fig. 6 is designed to carry the weeding implements, which can incorporate a 200 L tank for herbicide to enable multi-mode weed operations and even a small fertiliser spreader. And the front part of the implement unit houses the electronics and computers.

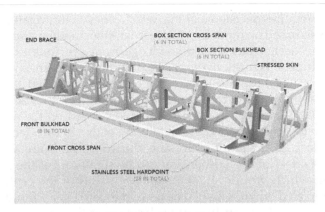

Fig. 6. Implement Unit.

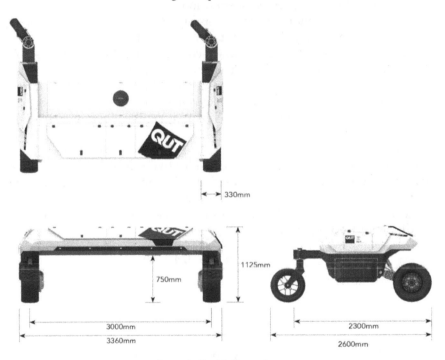

Fig. 7. AgBot II dimensions.

The main dimensions of the robot are summarised in Fig. 7. These are in agreement with Technical Specifications TS_{8-11} in Table 1. The width of implement unit is determined by a standard crop-row layout. The width of the side units and tyres are optimised with crop-inter-row spacing as a constraint. For an introduction to optimisation based design, the reader can refer to Arora (2004).

Design requires decision making, which relies on information obtained from analysis based on models, which can be either mathematical or mental models. The design based on knowledge from previous experiences— namely, induction is used. In the case of the chassis design, one important consideration was, the location of the centre of mass and the potential for vehicle to tip over, in case of an emergency brake. In order to analyse this potential, we built a mathematical model based on the assumptions that vehicle is considered to be a rigid body with a known location of the centre of mass. At the point of braking, the front wheels are assumed not to slip. The effect of the suspension can be significant if the suspension is not stiff, which would allow a significant shift of the centre of mass forward and also down. This can be accommodated in a simplified model by introducing offsets on the actual location of the centre of mass that represent the fully compressed suspension.

Consider the scenario depicted in Fig. 8. The vehicle is traversing down on a slope, the point of contact between the front wheel and the ground is P. As an approximation, we consider that this is also the point about which the vehicle will pivot during a sudden braking—the motion of the point is neglected and thus P is assumed to be fixed. The centre of mass is at the point C and the vehicle is assumed to have a mass m. At the point of braking, the vehicle traverses along the slope at a velocity $\vec{v}$ parallel to the slope.

Based on the modeling hypotheses, tipping occurs whenever point C moves forward of the point P. We consider two right-handed coordinate systems fixed to Earth frame, namely $\{0\}$ and $\{1\}$. As a generalised coordinate and first state variable, it is easy to take the angle θ of rotation about the y-axis (out of the page) of the line segment P-C. Then, the condition for tipping becomes

$$\theta > \frac{3}{2}\pi \quad (270°).$$

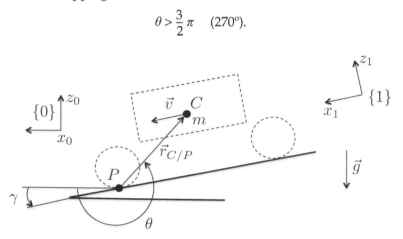

Fig. 8. Vehicle idealised physical system.

As a second state variable, we can choose the magnitude of the angular momentum of the centre of mass about the point P, namely $L = |\vec{L}|$. Then, a state-space model for system is

$$\dot{\theta} = J^{-1} L, \tag{1}$$

$$\dot{L} = mgl \cos \theta - b\, J^{-1} L, \tag{2}$$

where $l = |\vec{r}_{C/P}|$ is magnitude (distance) of the position of C relative to P, g is the acceleration of gravity, b is a damping coefficient, and $J = ml^2$ is the moment inertia of the centre of mass about P.

In order to simulate (1)-(2) and ascertain whether a vehicle may tip over, we set the initial conditions for θ and L based on the initial linear momentum of the vehicle and its operational speed. This simulation model was used to assess the risk of tipping given the location of the centre of mass. The model does not consider free-surface effects in the herbicide tank (which might alter the location of the centre of mass). It was also envisaged that the under-hanging weeding implement would contribute to lower the centre of mass thus reducing the risk of tipping relative to that provided by the model. So the model provides an idealised setting, and the decision as to whether the centre of mass is at the correct location has to be made under uncertainty, Tribus (1969).

4.2 Drive Unit and Power System

The main drive units for the AgBotII consist of a customised motor, gearbox and emergency brake assembly mounted inside a wheel hub. This is shown in Fig. 9. The design of this sub-system takes into consideration the vehicle drive and its power requirements, in conjunction with the torque, efficiency and load specification of the individual components.

A 5 kW, 48VDC electric motor, with an efficiency of 75–85% at 3200–4500 rpm, was paired with a 61:1 two-stage planetary gearbox to provide

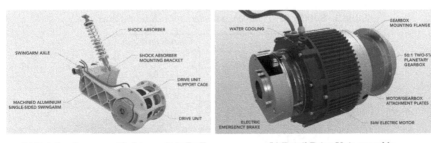

(a) Detail swing arm with drive unit AgBotII.   (b) Detail Drive Unit assembly.

Fig. 9. Swing arm and drive unit assembly.

energy efficient locomotion at the desired speed range of 5–10 km/h. The motor output shaft was re-designed to allow for the addition of a fail–safe electric brake, which is mounted directly to the rear face of the motor via a modified friction plate. The entire drive unit assembly is mounted to the vehicle's single-sided swing arm via a support cage, which transfers the load from the gearbox mounting flange to the swing arm.

The specification for the power-train follows from the vehicle's technical specifications given in Table 1. In particular, the total vehicle mass, maximum speed, maximum inclination angle and maximum acceleration determines the output power and torque of the power-train. Furthermore, the wheel diameter is a critical factor in the selection of a gear reduction ratio. Table 2 summarises the vehicle's specifications relevant to the power-train design.

One of the most critical factors affecting the selection of a power terrain is the rolling resistance coefficient. Agricultural vehicles operate in a wide variety of field conditions such as loose soil, compacted soil, paved roads and wet soil. Each of these surfaces have a different resistance coefficient shown in Table 3. The two coefficients of most interest are that of medium soil and the wet soil. The medium hard soil is the surface travelled on most often and has a steady-state power consumption. The wet soil is the worst surface the vehicle has to traverse and has the maximum power requirement.

Table 2. Key vehicle parameters related to power-train design.

Attribute	Magnitude	Units
Total Vehicle Mass	600	kg
Rated Speed	5	km/h
Maximum Speed	10	km/h
Acceleration	1	m/s$_2$
E-Brake Deceleration	4.9	m/s$_2$
Maximum Inclination	15	deg
Front Wheel Diameter	0.61	m

Table 3. Rolling Resistance Coefficients for different Surfaces.

Surface	Coefficient (C_r)
Smooth Concrete	0.01
Worn Asphalt	0.02
Gravel	0.02–0.03
Medium Soil	0.08
Loose Soil	0.1
Wet Soil, Mud	0.2
Sand	0.2–0.3

The mechanical power required to move the vehicle is given by:

$$P = P_{rolling} + P_{gradient} + P_{acceleration}$$
$$P = (C_r mg \cos \theta + mg \sin \theta + ma)v, \tag{3}$$

where C_r is the coefficient of rolling resistance, m is the total vehicle mass, θ is the maximum gradient, a is the maximum acceleration and v is the vehicle speed. The mechanical torque is given by the following similar equation:

$$\tau = (C_r mg \cos \theta + mg \sin \theta + ma)r, \tag{4}$$

where r is the radius of the drive wheels. We can calculate the steady-state power and torque using (3) and (4) equation with no gradient, no acceleration and a speed of 5 km/h on medium soil. The required power is 660 W and the required torque is 180 Nm. The peak power and torque requirements occur when the vehicle accelerates up to its rated speed on wet soil up a slope. That is, when $v = 5$ km/h; $a = 1.0$ m/s$_2$; $\theta = 15$ deg and $C_r = 0.2$. Under these conditions the required power increases significantly to 4.5 kW and the required torque increases to 1040 Nm.

The power efficiency of the drive unit is very important since AgBotII is powered by batteries. We can compute this efficiency as

$$\eta_u = \frac{P_o}{P_i} = \frac{T_u \omega_u}{V_u I_u}, \tag{5}$$

The output mechanical power P_o is given by the product of the torque T_u generated by the unit and the angular rate ω_u of the wheel. The input power P_i is given by the product of the voltage V_u and current I_u of the battery feeding the unit. The efficiency, thus, calculated includes the efficiency of the power electronics that control the motor, the efficiency of the electrical motor itself, and the efficiency of the gear box since Po is the power at the low speed side of the gear box. Figure 10(a) shows a rig used for testing the drive unit. The low speed shaft out of the gear box connects to the shaft of a hydraulic pump of known efficiency. By measuring the differential pressure and the volumetric flow of the hydraulic pump we can estimate the efficiency of the drive unit:

$$\eta_u = \frac{P D N \omega_m}{\eta_h V_u I_u}, \tag{6}$$

where P is the differential pressure at the hydraulic port, D is the displacement of the pump, N is the gear box ratio, and η_h is the known power efficiency of the pump. Figure 10(b) shows some of the lab tests conducted under different loads.

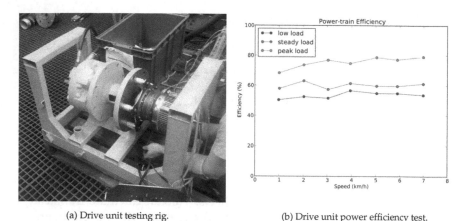

(a) Drive unit testing rig.   (b) Drive unit power efficiency test.

Fig. 10. Testing rig and drive unit test.

The final requirement of the power-train is the emergency braking torque. Emergency brakes apply a high constant torque to the wheels when they are engaged. We would like the emergency stopping distance of the vehicle to be as short as possible and for the wheels not to lock. Thus, we want the emergency brake torque to be as high as possible without locking the wheel. The vehicle will be required to operate on a wide range of surfaces with varying levels of available grip. At low speeds, the amount of available grip is characterised by the coefficient of friction, μ, between the tyre and the surface. The maximum deceleration on a surface with a given coefficient of friction is $a_{max} = \mu g$. Gravel roads, wet grass and soil have coefficients of friction of 0.35, 0.20 and 0.60 respectively, which correspond to maximum decelerations of 0.35 g, 0.2 g and 0.6 g, respectively.

As the emergency brake can only apply a single constant torque we need to use the lowest deceleration to avoid locking a wheel, that is, the target deceleration for an emergency stop will be 0.2g (2.0 m/s^2). If there is a single brake failure we still would like the vehicle to decelerate at 1.0 m/s^2 using the remaining brakes. Note that, it has been observed that during emergency braking a locked wheel has only a small effect on stopping distances on wet grass and actually decreases distances by 25% on deformable surfaces, such as gravel roads. Thus, if a wheel locks it should only have a small effect on the stopping distance.

The emergency braking (E-Brake) torque is given by:

$$\tau = (mg \sin \theta + ma - C_r mg \cos \theta)r, \qquad (7)$$

Table 4. Power–train Specifications.

Attribute	Magnitude	Units
Rated Power	0.66	kW
Rated Torque	180	Nm
Peak Power	4.5	kW
Peak Torque	1040	Nm
E-Brake Torque	875	Nm
Degraded E-Brake Torque	438	Nm

where m is the mass, a is the braking acceleration and r is the radius of the braking wheels. Note that, safety features ensure that the power-train is disabled in an emergency stop so that the brake will not work against the motors. When we use the above equation in the case of an emergency down a slope of 15 deg, the emergency torque was 875 Nm. To achieve the degraded emergency braking of 1.0 m/s², we need a torque of 438 Nm. Table 4 summarises the requirements for the power-train. The E-Brake torque places the greatest strain on the power-train and dictates the strength and size of the gearbox required. While the peak power requirement determines the size of the motor required.

4.3 Guidance, Navigation and Motion Control

Figure 11 shows a block diagram of the robot Guidance, Navigation, and Control (GNCC). In the following, we describe the attributes of the key subsystems.

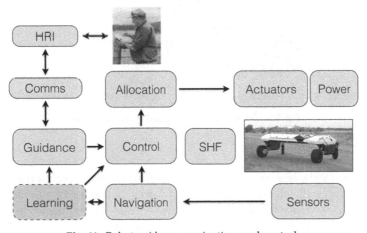

Fig. 11. Robot guidance, navigation, and control.

4.3.1 Navigation

The navigation system processes the data from the different motion sensors (RTK-GPS, IMU, wheel odometry) and fuses these data using observers to extract information about position and velocity of a particular point of reference on the robot and also the rate of turn about this point. The observers are built upon kinematic models that describe the geometric aspects of motion only.

4.3.2 Guidance

The mission planning of AgBotII sets the desired paths and triggers the turn and row-shift manoeuvres at the end of each crop row in the paddock. This is done through a user interface shown in Fig. 12(a).

Once the robot is on an active leg between two way points, the guidance system uses a standard line-of-sight guidance algorithm to provide a reference to the heading motion control system. In this algorithm the course to steer (or alternative the desired rate of turn) is computed based on the current position and heading so as the cross-track error—computed as the orthogonal projection of the point of reference on the robot onto the track. The configuration relevant to the line-of-sight guidance is shown in Fig. 12(b).

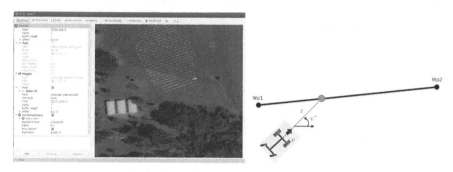

(a) User interface for operation planning.   (b) Line-of-sight configuration for guidance.

Fig. 12. Operation planning and guidance system.

4.3.3 Motion control

For mechanical systems, it is convenient to consider control as the function that generates the desired generalised forces in the particular degrees of freedom in which we seek to move the system. For example, in the case of AgBotII, we have two degrees of freedom: longitudinal translation and rotation about its vertical axis.

Based on the Lagrange-D'Alembert procedure for the formulation of equations of motion of non-holonomic systems, the following state-space model for the robot can be obtained:

$$\dot{u} = -\ell\omega^2 + \frac{1}{m}R(u, \omega) + \frac{1}{m}F_{D'} \tag{8}$$

$$\dot{\omega} = \frac{m\ell}{(I + m\ell^2)}\omega u + \frac{1}{(I + m\ell^2)}T_R(u, \omega) + \frac{1}{(I + m\ell^2)}T_{D'} \tag{9}$$

$$\dot{\psi} = \omega, \tag{10}$$

$$\dot{x} = u\cos(\psi), \tag{11}$$

$$\dot{y} = u\sin(\psi), \tag{12}$$

where u is the forward speed, ω is the rate of turn, ψ is the heading angle, and x and y are the local cartesian coordinates with respect to the point of reference O—see Fig. 13(a). The thrust force, denoted by F_D is produced by the drive train of the robot, and the rolling resistance force is denoted by $R(u, \omega)$. Similarly, in (9), we have added the driving torque T_D produced by the differential steering of the drive train, and the resistance friction torque $T_D(u, \omega)$. The parameter m is the mass of the robot, I is the moment of inertia about its centre of mass, and ℓ is the offset of the centre of mass C relative to the point of reference P used to formulate the equations of motion. Details of the configuration space are depicted in Fig. 13(a).

Figure 13(b) shows a block diagram of a motion control system that is used in AgBotII. The speed controller takes the reference speed v^* and the actual velocity v estimated by the navigation system, and provides a set point of thrust force τ_1. The heading controller has two nested feedback loops. The inner loop is an angular rate loop and the outer loop is the heading angle loop. The heading loop uses information about the desired heading ψ^* and the actual heading ψ to generate a reference angular rate

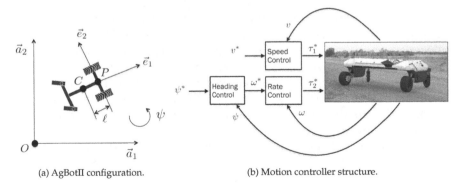

(a) AgBotII configuration.   (b) Motion controller structure.

Fig. 13. AgBotIIconfiguration space and motion control structure.

ω^* for the inner control loop. The latter uses this information together with the actual angular rate ω to generate the desired control torque τ_2^* in the vehicle. Both, the speed and rate controllers have an integral action. In order to design the controllers, we can linearise the rate state-space model about the point $u = \bar{u}$ given by the nominal speed and $\omega = \bar{\omega} = 0$. This, decouples the model for design, but the robustness of the controller should be assessed in terms of the coupled nonlinear model. As we can see, a key advantage of the actuator configuration of AgBotII, from the point of view of control design, is that the two degrees of freedom of interest can be decoupled for control design. This simplifies the control design task. The control structure depicted in Fig. 13(b) has been successfully applied to marine and aerospace vehicles.

Note that once the vehicle reaches the end of the paddock and requires turning, the structure of the control system remains the same; it is the guidance system which switches modes.

Vehicles with wheels are subject to non-holonomic constraints; this means that although the vehicle can take any position and orientation on the plane (a configuration), the trajectories that take the vehicle from one configuration to another are restricted. For example, AgBotII cannot move sideways, but using a combination of forward, backward and turning motion it can be at a pose to the side of its initial pose. The guidance system must address the problem of generating feasible trajectories in agreements with the vehicle motion constraints.

4.4 Control Allocation

The use of control allocation is borrowed from aerospace and marine vehicles. In general, it provides two key features to motion design of vehicles. First, the motion controller is designed to output forces; and hence, the design of this controller is kept within the realm of mechanics. This has an advantage that concepts related to energy and passivity can be used in the design and demonstrate stability properties of the controller. Second, in systems that are overactuated (not the case of AgBotII though), control allocation provides a way of implementing tolerance to actuator failures, since, it can shut down a failed actuator and re-configure the remaining healthy ones to provide the desired control forces. This does not require re-tuning the controller or switching to a different set of control gains.

In order to describe the control allocation mapping, we need to understand how the actuators produce the generalised forces on the robot.

Each traction wheel of the robot is commanded by and electrical motor coupled to the wheel with a gearbox of ratio $N > 0$. If we assume that all the moments of inertia of the motor rotors and the wheels have been reflected and are lumped in the mass m of the robot, then we can consider each gearbox and wheel as the combination of two ideal transformers—thus, preserving power. For each gearbox, we then have the following balance of power:

$$T_m \omega_m = T_w \omega_w, \tag{13}$$

where T_m and ω_m are the motor torque and angular rate and T_w and ω_w are the wheel torque and angular rate. Hence, the torque on the wheel can be written as a function of the torque on the motor:

$$T_\omega = \frac{\omega_m}{\omega_w} T_m = N T_m, \quad \text{where, } N = \frac{\omega_m}{\omega_w} \tag{14}$$

For the ideal wheel (massless) of radius r,

$$T_w \omega_w = F_w u_w, \tag{15}$$

where F_w is the thrust produced by the wheel and u_w is the linear translational velocity of its axle. Hence,

$$F_w = \frac{\omega_w}{u_w} T_w = \frac{1}{r} T_w, \quad \text{where, } r = \frac{u_w}{\omega_w} \tag{16}$$

Therefore, the combined relationship between the torque and the thrust of the wheel is

$$F_w = \frac{N}{r} T_m. \tag{17}$$

The total driving force on the robot is the sum of thrust of the left and right wheel (F_L, F_R) namely,

$$F_D = F_L + F_R = \frac{N}{r}(T_{Lm} + T_{Rm}). \tag{18}$$

where, T_{Lm}, T_{Rm} are the thrust on left and right wheel.

The torque due to the differential driving is given by

$$T_D = -dF_L + dF_R = \frac{dN}{r}(-T_{Lm} + T_{Rm}), \tag{19}$$

where d is the offset of the wheel from the vehicle's centre line. Expressions (18) and (19) provide the inputs to the state-space model (8)–(12).

The control allocation function maps the desired generalised forces τ^* that the motion controller demands into desired actuator commands δ^*:

$$\delta^* = f_c(\tau^*). \tag{20}$$

These actuator commands (voltage, currents, etc.) are the input to the local actuator controllers that implement the true actuator command δ.

The latter is then mapped into the actual generalised forces τ by the actuator configuration mapping:

$$\tau = f_A(\delta). \tag{21}$$

Under perfect control allocation the composition $f_A \circ f_C$ should be the identity mapping, and thus $\tau^* \mapsto \tau$.

In the case of AgBotII, we consider the following vectors:

$$\tau \triangleq \begin{bmatrix} F_D \\ T_D \end{bmatrix}, \qquad \delta \triangleq \begin{bmatrix} T_{Lm} \\ T_{Rm} \end{bmatrix}.$$

Then, the actuator configuration mapping $f_A: \delta \mapsto \tau$ can be expressed as

$$\begin{bmatrix} F_D \\ T_D \end{bmatrix} = \frac{N}{r} \begin{bmatrix} 1 & 1 \\ -d & 1 \end{bmatrix} \begin{bmatrix} T_{Lm} \\ T_{Rm} \end{bmatrix}$$

and the control allocation mapping $f_C: \tau^* \mapsto \delta^*$ can be expressed as

$$\begin{bmatrix} T^*_{Lm} \\ T^*_{Rm} \end{bmatrix} = \frac{r}{N} \begin{bmatrix} 1 & 1 \\ -d & 1 \end{bmatrix}^{-1} \begin{bmatrix} F^*_D \\ T^*_D \end{bmatrix}$$

Due to the integral action of the speed and rate controllers, these should be implemented within a multivariable anti-wind up scheme. Such implementation is beyond the scope of this chapter.

5. WEEDING IMPLEMENT DESIGN

The integrated multi-mode weed destruction system for AgBotII incorporates both a selective mechanical weeding system and selective spray system. Individual weed species can be targeted by either the mechanical or spray system depending on the result of the vision-based weed detection and classification system.

To identify weeds, we developed a vision-based on-board detection and classification system comprising of a ground facing camera and an image processing computer. Weeds are first detected using colour information in the image and then an algorithm determines the species. The prototype assembly for the weed detection and classification system is capable of working continuously throughout the night and during daylight periods of uniform lighting.

Once the weeds are detected and classified, the system determines the best method for weed destruction and triggers the actuation of either the mechanical or spray modules. The modules are attached to the underside of the AgBotII platform and for the purposes of the first prototype have

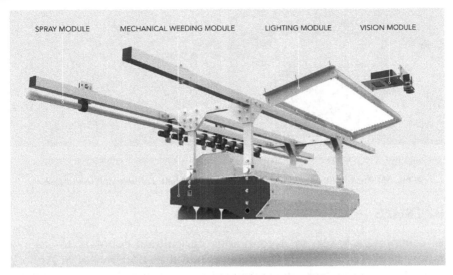

SPRAY MODULE   MECHANICAL WEEDING MODULE   LIGHTING MODULE   VISION MODULE

Fig. 14. Weeding implement system modules.

been designed to a width of 1 m. This can be extended to the full width of the robot in future designs. An overview of the weeding system attached to the AgBotII can be seen in Fig. 14.

The key aspects of the design in mechatronic are the inter-relations amongst:

- Processing speed of the vision-based weed detection and species classification system;
- Accuracy of the navigation and motion control system;
- The width of the individual mechanical implement and the speed of actuation.

The latency due to the processing speed of the vision-based weed detection and species classification system combined with the operational speed of the robot is used to determine the spatial separation between the camera and the weeding actuator and the speed of actuation. Whereas the number of actuators and the expeted number of actuators active determine the energy storage requirements of the robot as well the size of the accumulator of the pneumatic system that activate the mechanical weeding implements.

In order to assess these aspects of the design, we constructed a simulation model—shown in Fig. 15. This model together with a probabilistic model about the weed density was used in an iterative design to determine the dimensions and the speed of actuation of the weeding array.

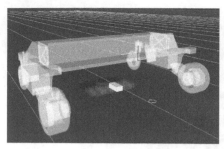 

(a) Simulated weed camera and weed implements.   (b) Simulation of entire weed system.

Fig. 15. Simulator used to assess design characteristics of weeding operations.

6. TRIALS

A common task for an autonomous agricultural robot is to keep a fallow field free of weeds. To demonstrate the effectiveness of AgBotII in performing this task, an experiment was conducted on a small fallow field. AgBotII was run over the field twice per week and the weeds were removed using the mechanical weeding implement as they appeared.

A scale map of the field used for the trial is shown in Fig. 16, the field was approximately 1000 m² (26 m by 38 m). The field was initially ploughed using a large tractor and then left alone to allow the weeds to emerge naturally. The field had a very large seed bank ensuring the growth of a large number of weeds when it rains.

The field was divided into one metre wide rows as shown in Fig. 16. One third of the field was selected to remain untouched as a control for the experiment, while the remaining two thirds had the weeds removed by the AgBotII. The rows to be used as control rows were selected at random and their distribution is also shown in Fig. 16.

The coverage of weeds (density per unit of area) in the control and treated rows were recorded on each iteration over 42 days. The amount of the soil tilled by the mechanical implements was also recorded on each iteration. There was a significant rain early in the experiment that caused a very large number of weeds to germinate at the same time; between day 17 and 24 the weed density in the field increased from 0.12 weeds per m² to 37.3 weeds per m².

Figure 17(a) shows the amount of weed coverage and the response of the autonomous weeding array throughout the field test. Figure 17(b) displays the percentage of weed coverage in the treated and control sections of the field. It is clear from the figure that after the rain, the amount of weed coverage in the control grows rapidly, however, the coverage in the treated area grows slowly and then starts to recede. The weed coverage in the treated area peaks at 4.5% while the weed coverage

Fig. 16. Field used for mechanical weeding trials. Clear colour areas correspond to robotic weeding areas.

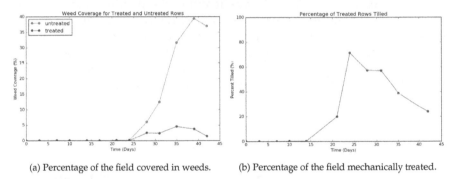

(a) Percentage of the field covered in weeds.   (b) Percentage of the field mechanically treated.

Fig. 17. Weed growth and subsequent weed treatment over time during the trial.

in the control area reaches 37%. At the conclusion of the trial, the weed coverage in the treated area is reduced to 1.5% demonstrating the efficacy of the AgBotII and the autonomous weeding array at weed management in a fallow field. The weed coverage is non-zero because new weeds are continuously germinating in the field due to the large seed bank and the summer weather.

Another way to measure weed coverage is to count the number of weeds. In the initial stages of the trial this was possible. However, as the weed density increased and the weeds grew larger counting was no longer possible in the control areas. Thus, the percentage area of weed coverage is used as a measure for the amount of weeds in the field. In the treated sections of the field the weed density peaked at 11.3 weeds per m^2 on day 21.

We also conducted a showcase on the 7th December 2016 at "Glenwoon", (Bowenville, Queensland). The system was transported to the property two days in advance of the showcase to allow time to set-up and trial the systems performance in a broadacre farm environment. Two demonstrations were organised and were run throughout the day. The morning session was attended local farmers and representatives from the advanced manufacturing sector. Figure 18 shows some of the group members attending the morning session demonstrating the robot capabilities.

The feedback from the attendees was overwhelmingly positive, with many of the farmers impressed with AgBotII weed detection and mechanical removal capabilities. Farmers were most interested in the potential for the vision and mechanical weeding capabilities. Many of the farmers were also interested in the plans to market this technology, and were keen on being part of any future industrial trials and testing.

A particular situation of interest developed in both morning and afternoon sessions. As AgBotII was roaming the paddock mechanically weeding, a few farmers asked *"Why is it activating the weeding implement when there are no weeds?"* Upon inspecting the soil where the robot had been weeding, we found small weeds 5 mm to 10 mm diameter leaves and 20 mm root. The farmers were very impressed. One farmer commented that if he sends people to chip weeds, they would rarely pick that.

Fig. 18. Demo trials conducted for farmers.

7. CONCLUSION

In this chapter, we discuss key aspects of mechatronic design for a weed-management robotic system. We highlight the process going from functional and operational specifications to technical specifications under which a design is to be conducted and feasible designs are to be assessed. We also attempted to describe the key interactions among design decisions in mechanical components, dynamics, hardware, actuators and control. These design interactions are at the *raison d'être* of a mechatronics approach to design. Then, we proceed to specify various aspects of the design and the factors leading to particular decisions about the design of our prototype robot. We finish the chapter with the results of a six-week trial on mechanical weeding.

REFERENCES

Arora, J. 2004. Introduction to optimum design. Elsevier Academic Press, second edition.

Bawden, O. J. 2015. Design of a lightweight, modular robotic vehicle for the sustainable intensification of broadacre agriculture. Master's thesis, Queensland University of Technology, Australia.

Gulliksen, J., Göransson, B., Boivie, I., Blomkvist, S., Persson, J., and Cajander, Å. 2003. Key principles for user-centred systems design. Behaviour & Information Technology. 22(6): pp. 397–409.

NTC. 2015. Australian Light Vehicle Standards Rules 2015. Technical report, National Transport Commission, Australia.

Redhead, F., Snow, S., Vyas, D., Bawden, O., Russell, R., Perez, T., and Brereton, M. 2015. Bringing the farmer perspective to agricultural robots. In Proceedings of Conference on Human Factors in Computing Systems, pp. 1067–1072.

Tribus, M. 1969. Rational descriptions, decisions, and designs. Pergamon unified engineering series: engineering design section. Pergamon Press.

Upadhyaya, M., and Blackshaw, R. 2007. Non-chemical Weed Management: Principles, Concepts and Technology. CAB International.

3

Robotics for Spatially and Temporally Unstructured Agricultural Environments

Konrad Ahlin,[1] Brad Bazemore,[2] Byron Boots,[1] John Burnham,[2] Frank Dellaert,[1] Jing Dong,[1] Ai-Ping Hu,[3,] Benjamin Joffe,[3] Gary McMurray,[3] Glen Rains[2] and Nader Sadegh[1]*

1. INTRODUCTION

The farming industry faces many problems that threaten its sustainability. Among the most important are the detection, prevention, and control of devastating plant pests and diseases. Management of pests and diseases (in addition to water and nutrients) is based on scouting a field weekly. Farmers can spend between 10 to 40 US dollars per acre to have their fields inspected by a human field scout. Depending on the crop, detection of even a single insect can trigger an intensive pesticide spraying program. On the other hand, non-comprehensive scouting may miss populations of pests that sometimes congregate in localized areas or a disease that is asymptomatic until it has spread to many areas of the field. As a consequence of this inefficient and sometimes inaccurate method, farmers spray preventatively for many plant pathogens. If more extensive and efficient quantification of pest control, water stress, and nutrient needs were possible, a tremendous cost savings could be achieved by a decrease in unnecessary spraying. Currently, the only solution to potential pest problems is to spray at the first sight of pests and treat the entire field. This leads to an overuse of pesticide which is costly and environmentally

[1] Georgia Institute of Technology, Atlanta, Georgia, USA.
[2] University of Georgia, Tifton, Georgia, USA.
[3] Georgia Tech Research Institute, Atlanta, Georgia, USA.
* Corresponding author: ai-ping.hu@gtri.gatech.edu

unfriendly. However, the risk of losing crops to pests is too high not to take necessary precautions.

The development of an automated field scout (AFS) would make it possible to determine the spatial and temporal distribution of pests and diseases, nutrient deficiencies, and water stress in a field. The AFS would help the farmer assess management strategies after they are implemented and help determine best management practices. This work describes a collaborative project among the Georgia Tech Research Institute, the Georgia Institute of Technology, and the University of Georgia in developing and fielding an AFS system composed of four main components: an autonomous ground vehicle, a vehicle-mounted 4-dimensional (4D) mapping system, a vehicle-mounted robot arm used for leaf and soil sampling, and a farmer/ consultant who will interact with the AFS system to meet the needs of each particular farm. The project focuses on peanuts, though the developed AFS could be adapted for any crop that requires intensive management.

In the following sections, the autonomous ground vehicle (referred to as the Red Rover) is first described, followed by a consideration of 4D mapping for agricultural applications. Robot arm control using visual servoing is then discussed for two particular scenarios: leaf sampling and apple picking (included here for its novel use of dual robot arms). Implementation results are described in their respective sections.

2. AUTONOMOUS GROUND VEHICLE

There are a few autonomous systems that have been developed around the world, and at least one is in commercial use. The Autonomous Tractor Corporation (Fargo, ND) offers a platform to connect multiple implements to their modular systems. This system is made specifically for planting seeds, but has been re-purposed for row crop inspection. Kinze manufacturing is testing a driverless tractor that pulls a grain cart in tandem with the grain combine for unloading during harvest. John Deere (Machine Sync) and CaseIH (V2V System) are developing similar systems but with a driver in the seat. The combine takes control of the tractor pulling the grain cart and matches speed and distance to unload automatically, reducing stress and need for highly experienced drivers.

While companies like Google may be garnering all the publicity for autonomous vehicles, research has been robustly moving forward in development of operator-assisted and fully autonomous systems in agriculture. GPS-guided and steered tractors have shown incredible benefit in reduced planting and harvesting losses (Bergtold, 2009; Vellidis et al., 2013), reducing driver fatigue, and extending work hours in low visibility. Complete autonomy remains elusive due to safety concerns, but the agricultural industry is moving in that direction. Autonomous vehicles for orchard operations in citrus have been proven (Subramanian, 2009).

The use of unmanned aerial systems (UAS) has gained a lot of attention for scouting fields. However, it appears that any detection of plant stress from the air will require ground-truthing to isolate the cause and come up with a management strategy for the crop. The AFS would provide persistent and more comprehensive coverage of the field for this crucial ground-truthing.

2.1 Hardware

The Red Rover AFS (Fig. 1) is a custom-built articulated vehicle (West Texas Lee Corp.) with modifications to meet the harsh environmental, navigation, and obstacle avoidance requirements of a field that has both unstructured (open areas and end of rows) and structured elements (crops in rows are visible). The drive system is hydrostatic and the left and right turns are achieved using hydraulic actuators powered by a 0.45 cc/rev fixed displacement pump (Bucher Hydraulics, Italy) and a 4-port, 3-way closed center solenoid actuated DCV. The fixed gear pump is connected in tandem with an axial-piston variable-rate pump (OilGear, maximum displacement of 14.1 cc/rev) with swashplate for directing hydraulic fluid for forward and rearward movement, to all four wheels. The pump tandem is powered by the onboard Kohler 20 HP gasoline engine. The swashplate angle of the hydrostatic drive is controlled by a microcontroller using a

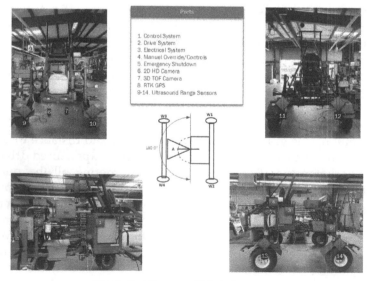

Fig. 1. Red Rover and labeled systems.

20 cm stroke electric linear servo. Each hydraulic wheel motor has a 240 cc/rev hydraulic motor (Parker Hydraulics, USA). Maximum articulation angle is 45 degrees and the wheelbase is 180 cm. Height and width of the vehicle can be adjusted up to a maximum clearance of 122 cm and a maximum width of 234 cm.

A waterproof 90-degree rotation servo motor and microcontroller manages the engine RPM via the engine throttle. Engine RPM will be monitored for safety as well as for control of the speed and turning rate of the Red Rover.

There are 2 electrical boxes on the rover. One holds the main server or Robot Operating System (ROS) Master and the other contains relays, microcontrollers, and two 12V DC to DC voltage converters. A forced air cooling system is used to keep each of these boxes within specific temperature limits (wide fluctuations can warp the electrical boards). A temperature feedback system was developed to pump air into the vortex tube and deliver it to the electrical boxes. A compressor and air tank is used to keep compressed air available at 100 psi and 10 cubic feet per minute.

To protect the Red Rover and its electronic systems, as well as pedestrians working in the vicinity of its operation, obstacle avoidance components and electrical overload protection have been incorporated. For example, in case of a detected obstacle, the controller will automatically send the swashplate to neutral and reduce engine RPM. If collision continues to be imminent, the system will shut itself off. This is accomplished using a combination of RGB camera, time-of-flight camera, and ultrasonic sensors to provide the Red Rover awareness of its surroundings. Each sensor represents a node that passes information to the ROS Master and data is processed using a Kalman filter to assess obstacles and assist GNSS/IMU navigation. Information from the filter will be passed back to ROS to ascertain if a controlled safety maneuvers is needed.

RTK-GPS is being integrated using the Piksi SWIFT navigation system (Swift Navigation, USA). This system is low-cost and highly accurate. Currently one receiver is being used as the base station and the other as the Rover. Future development will incorporate a cellular modem and NTRIP network communications from a continuously operating reference station. Rover navigation utilizes ROS and user-defined path-following algorithms that incorporate the multiple awareness sensors (Rains et al., 2014).

2.2 Software

There are three main software components for control and navigation of the Red Rover. All 2D visual data captured through a RGB camera will be processed by OpenCV machine vision algorithms for object detection and identification for obstacle avoidance, as well as crop identification for support of vehicle navigation. 3D depth of field data collected by the time-of-flight camera will be processed by the OpenCV Point Cloud Library. The two data streams, along with GNSS data, will be given to ROS for mapping and control of the Rover's maneuvers.

For further navigation from one location in the field to another, the paths of each crop row will already be stored on-board and an optimization routine is used to find the shortest path to relocate for data collection. Specific locations will be relayed manually by the farmer or crop consultant based on analysis of aerial data, ground data, and historical experience. Rover control and communications rely on Ethernet protocols and are integrated using ROS libraries. Ethernet is a standard communications protocol and a gigabit main switch is used to provide the communications speed required for real-time sensing and feedback controls. Diagnostic sensors (engine temperature, engine RPM, and hydraulic pressure) and navigation and avoidance sensors (GNSS, IMU, cameras) will each be defined as nodes within the ROS architecture. Remote control and observance of the system properties are available through a web-server on the ROS master and a cellular data modem. Control and observation of the on-board sensors will be accessible over multiple outlets for the farmer/consultant using the system. The on-board GUI will be available from the data modem on remote terminal, tablet, or smartphone interface. From any of these, a user will have access to the ROS nodes and sensor data from any of the cameras, GNSS, robot arm, and diagnostic sensors. This adds a level of observing ability, that may be beneficial for examining plants in real-time from a remote location or in diagnosing problems with the Red Rover system operations.

3. 4D MAPPING

Computer vision is a powerful tool for monitoring crops and estimating yields with low-cost image sensors (Hague et al., 2006; Nuske et al., 2014; Font et al., 2015; Sa et al., 2016). However, the majority of this work only utilizes 2D information in individual images, failing to recover the 3D geometric information from sequences of images. Structure from Motion (SfM) (Agarwal et al., 2009) is a mature discipline within the computer vision community that enables the recovery of 3D geometric information from images. When combined with Multi-View Stereo (MVS) approaches (Furukawa and Ponce, 2010), these methods can be used to obtain dense, fine-grained 3D reconstructions. The major barrier to the direct use of these methods for crop monitoring is that traditional SfM and MVS methods only work for static scenes, which cannot solve 3D reconstruction problems with dynamically growing crops.

We address here the problem of time-lapse 3D reconstruction with dynamic scenes, to model continuously growing crops. We call the 3D reconstruction problem with temporal information, 4D reconstruction. The output of 4D reconstruction is a set of 3D entities (point, mesh, etc.), associated with a particular time or range of times. An example is shown in Fig. 2. A 4D model contains all of the information of a 3D model, e.g., canopy size, height, leaf color, etc., but also contains additional temporal information, e.g., growth rate and leaf color transition. We also collected a field dataset using a ground vehicle equipped with various sensors, which we will make publicly available. To our knowledge, this will be first freely available dataset that contains large quantities of spatio-temporal data for robotics applications targeting precision agriculture.

June 9

June 23

Aug 1

Fig. 2. Reconstructed 4D model of a peanut field by our approach. Each time slice shown has been reconstructed from a dense point cloud.

We cite three main contributions of our 4D mapping research:

- Development of an approach of 4D reconstruction for fields with continuously changing scenes, mainly targeting crop monitoring applications.
- Development a robust data association algorithm for images with highly duplicated structures and significant appearance changes.
- We collect a dataset containing ground truth crop statistics obtained from a field vehicle for evaluating 4D reconstruction and crop monitoring algorithms.

We begin by stating several assumptions related to crop monitoring, before specifying the details of our 4D reconstruction algorithm.

- The scene is static during each data collection session.
- The field may contain multiple rows.

The first assumption is acceptable because we only focus on modeling crops and ignore other dynamic objects like humans. The crop growth is also too slow to be noticeable during a single collection session. The second assumption is based on the geometric structure of a typical field. The 4D field model reflecting these two assumptions is illustrated in Fig. 3.

Our proposed system has three parts.

1. A multi-sensor Simultaneous Localization and Mapping (SLAM) pipeline, used to compute camera poses and field structure for a single row in a single session.
2. A data association approach to build visual correspondences between different rows and sessions.
3. An optimization-based approach to build the full 4D reconstruction across all rows and all sessions.

To generate the 4D reconstruction of the entire field, we first compute 3D reconstruction results for each row at each time session, by running

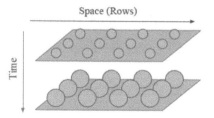

Fig. 3. The field 4D model. The field contains multiple rows and there are multiple time sessions of the field.

multi-sensor SLAM independently. Next, we use the data association approach to match images from different rows and sessions, building a joint factor graph that connects the individual SLAM results. Finally we optimize the resultant joint factor graph to generate the full 4D results.

3.1 Multi-Sensor SLAM

The SLAM pipeline used in this work has two parts, illustrated in Fig. 4. The first part of the SLAM system is a front-end module to process images for visual landmarks. SIFT (Lowe, 2004) features are extracted from each image and SIFT descriptor pairs in nearby image pairs are matched using the approximate nearest neighbor library FLANN (Muja and Lowe, 2014). The matches are further filtered by 8-point RANSAC (Hartley and Zisserman, 2004) to reject outliers. Finally, a single visual landmark is accepted if there are more than 6 images that have corresponding features matched to the same landmark.

The second part of the SLAM system is a back-end module for estimating camera states and landmarks using visual landmark information from the front-end and other sensor inputs. Since the goal of the multi-sensor SLAM system is to reconstruct a single row during a single data collection session, the back-end module of the SLAM system estimates a set of N camera states X, at row r_i and time t_i, given visual landmark measurements from the front-end module, and other sensor measurements, including an Inertial Measurement Unit (IMU) and GPS.

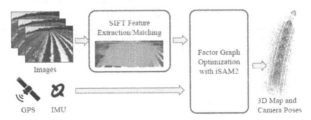

Fig. 4. Overview of multi-sensor SLAM system.

3.2 Robust Data Association over Time and Large Baseline

The second key element of our approach is robust data association. Data association is a key technique to get reconstruction results of more than a single row at a single time; however, the data association problem between different rows or times is difficult, since there are significant appearance changes due to illumination, weather or view point changes. The problem is even more difficult in crop monitoring due to measurement aliasing (Indelman et al., 2016): fields contain highly periodic structures with little visual difference between plants (see Fig. 2). As a result, data association problems between different rows and times is nearly impossible to solve by image-only approaches.

Rather than trying to build an image-only approach, we use single row reconstruction results output by SLAM as a starting point for data association across rows and time. The SLAM results provide camera pose and field structure information from all sensors (not just images), which helps us to improve the robustness of data association.

3.3 4D Reconstruction

The third and the last part of our pipeline is a 4D reconstruction module. The complete 4D reconstruction pipeline is illustrated in Fig. 5. We define the goal of 4D optimization as jointly estimating all camera states X and all landmarks L across all rows and times (sessions). The measurements Z includes all single row information as well as data association measurements Z_{cr} that connect rows across space and time.

Data association is performed across different rows and times to get Z_{cr}. Exhaustive search between all row pairs is not necessary, since distant rows are not visible from each other in the images, and large timespans makes matches between images difficult to calculate. In our approach we only match rows next to each other in either the space domain (nearby rows in the field), or the time domain (nearby date).

The point cloud estimate of L is relatively sparse, since it comes from a feature-base SLAM pipeline, where only points with distinct appearance are accepted as landmarks (in our system SIFT key points are accepted). An optional solution is to use PMVS (Furukawa and Ponce, 2010), which takes estimated camera states to reconstruct dense point clouds.

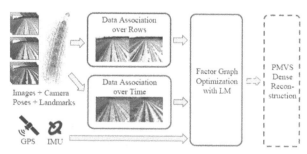

Fig. 5. Overview of 4D reconstruction pipeline. Dash box of PMVS dense reconstruction step means it is optional.

3.4 Dataset

To evaluate the performance of our approach with real world data, we collected a field dataset with large spatial and temporal scales. Existing datasets with both large scale spatial and temporal information include the CMU dataset (Badino et al., 2011), the MIT dataset (Fallon et al., 2012), and the UMich dataset (Carlevaris-Bianco et al., 2016). However, all of these datasets are collected in urban environments, and are not suitable for precision agriculture applications.

The dataset was collected from a field located in Tifton, GA, USA. The size of the field is about 150 m × 120 m, and it contains total 21 rows of peanut plants. The map of the field is shown in Fig. 6. We use a ground vehicle (tractor) equipped with multiple sensors, shown in Fig. 6, to collect all of the sensor data. The equipped sensors include: (1) a Point Grey monocular global shutter camera, 1280 × 960 color images are streamed at 7.5 Hz, (2) a 9DoF IMU with compass, acceleration and angular rate are streamed at 167 Hz, and magnetic field data is streamed at 110 Hz, (3) a high accuracy RTK-GPS, and a low accuracy GPS, both of them stream latitude and longitude data at 5 Hz. No hardware synchronization is used. All data are stored in a SSD by an on-board computer.

We recorded a complete season of peanut growth which started May 25, 2016 and completed Aug 22, 2016, right before harvest. The data collection had a total of 23 sessions over 89 days, approximately two per week, with a few exceptions due to severe weather. Example images of different dates are shown in Fig. 7. Each session lasted about 40 minutes, and consisted of a tractor driving about 3.8 km in the field.

Fig. 6. Top left is the tractor collecting the dataset; Down left shows sensors and computer (RTK-GPS is not shown); Right is a sample RTK-GPS trajectory, and sites of manual measurements are taken, overlay on Google Maps.

Fig. 7. Eight sample images taken at approximately same location in the field, dates are marked on images.

In addition to sensor data, ground truth crop properties (height and leaf chlorophyll) at multiple sampling sites in the field were measured weekly by a human operator. There were a total of 47 measuring sites, as shown in Fig. 6.

3.5 Results

We ran the proposed 4D reconstruction approach on the peanut field dataset. We implemented the proposed approach with the GTSAM C++ library. We used RTK-GPS data from the dataset as GPS input, and ignored lower accuracy GPS data. Since the peanut field contains two sub-fields with little overlap (see Fig. 6), the two sub-fields were reconstructed independently and aligned by GPS. Since the tractor runs back and forth in the field, we only use rows in which the tractor driving south (odd rows), to avoid misalignment with reconstruction results from even rows. An example of densely reconstructed 4D results shown in Fig. 2.

Although Fig. 2 shows that the 3D reconstruction results for each single session qualitatively appear accurate, to make these results useful to precision agriculture applications, to precision agriculture applications they should be evaluated quantitatively as well. In particular we wanted to answer the following questions:

- Are these 3D results correctly aligned in space?
- Are these 3D results useful for measuring geometric properties of plants, useful for crop monitoring (height, width, etc.)?

To answer the first question, we visualize the 4D model by showing all 3D point clouds together. We visualize part of the 4D sparse reconstruction, result is shown in Fig. 8. Point clouds from different dates are marked in different colors. We can see from the cross section that the ground surface point clouds from different sessions are well aligned, which shows that all of the 3D point clouds from different dates are well registered into a single coordinate frame. This suggests that we are building a true 4D result. We can see the growth of the peanut plants, as the point cloud shows

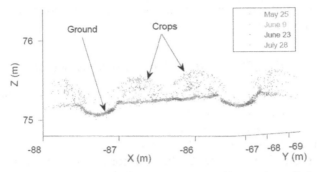

Fig. 8. Cross-section of part of the sparse 4D reconstruction results at 3rd row. Only 4 sessions are shown to keep figure clear.

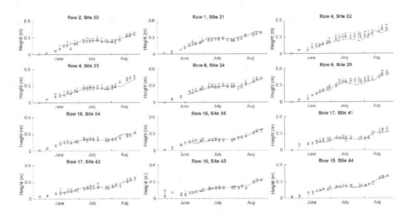

Fig. 9. Estimated peanut heights at 12 sampling sites in blue, with ground truth manual measurements in black lines.

"Matryoshka doll" like structure, earlier crop point clouds are contained within point clouds of later sessions.

To answer the second question, we show some preliminary crop analysis results using reconstructed 4D points and compare them to ground truth measurements we took manually. We setup a simple pipeline to estimate height of peanut plants from sparse reconstructed 4D point clouds at multiple sites, by first estimating the local ground plane by RANSAC from May 25's point cloud (when peanuts are small and ground plane is well reconstructed), second separate peanut's point clouds by color (using RGB values), and finally estimate the distance from peanut canopy's top to ground plane. Preliminary height estimations of twelve sampling sites are shown in Fig. 9. With the exception of sites 22 and 25, which have slightly biased estimated heights due to poor RANSAC ground plane estimations, results of the sites meet the ground truth measurements well. This shows that we can compute reasonable height estimates even with a simple method, and proves that the 4D reconstruction results contain correct geometric statistics.

4. ROBOT ARM CONTROL USING VISUAL SERVOING

Robot arms work best in structured environments where their desired tasks are known before they need to be executed. However, the field of robotics is ever expanding out of factory-like environments and into broader applications. Thus, the demand for completing unstructured tasks, tasks that have elements that cannot be known ahead of time, is increasing. This research will focus on the unstructured task of having a robotic arm grasping a leaf from a plant. The challenges in this task involve both the identification of a leaf within the space of a camera image as well as locating the leaf in the 3D Cartesian world. While the type of leaf can be assumed to be known, and the approximate location of the plant is assumed to be known, neither the exact geometry of the leaf nor the approximate location of the leaf is known. The task for the arm is to search for the leaf, identify the leaf, and then, using 2D pixel information, grab the leaf with its manipulator.

Identifying and localizing a leaf is not a trivial problem. Leaves vary in size and appearance, and are susceptible to overlapping and occlusions. Moreover, the implementation should be robust to variations in natural illumination. Deep learning techniques have achieved great results in object detection, while demonstrating good computational performance by using modern GPU computing. Once information about the leaf can be determined in the 2D image space, the manipulator can attempt a routine to grasp the object.

Visual servoing is the process of controlling a robotic device using real-time visual information in a feedback loop. Image Based Visual Servoing (IBVS) is a classical approach to visual servoing in robotics that attempts to converge on an object in 3D space by only using information about the objects 2D pixel geometry. IBVS does not assume any information about the object being viewed; instead, this method uses a given set of desired points in the image space that it uses for its error calculations (Chaumette and Hutchinson, 2006). IBVS is a widely studied topic that is often used in robotics applications.

Robotic arms typically have a virtue in their design that offer an alternative or supplement to the IBVS approach. Assuming that the kinematics of the arm and the rotation of the joint angles are known (a necessity for most joint control methods), the final position of the end-effector can be determined. Since the camera is rigidly attached to the manipulator, the position and orientation of the camera is always known in Cartesian space. Using the location of the camera and the pixel coordinates of desirable feature points, the 3D location of images can be determined. Monoscopic Depth Analysis (MDA) uses multiple images and triangulation to deduce the position of feature points in Cartesian space relative to the camera. Although the method of visual triangulation is widely known, the implications of using the MDA approach for robotic manipulators has a vast potential. MDA is a simple routine that has been used in other visual applications (e.g., Kalghatgi and Sadegh, 2012) and holds promise in improving the control of robotic manipulators.

The task of picking a leaf can be separated into the following, unstructured sub-tasks: determining the position of the leaf in the image space and transforming its position into Cartesian space. The method of using Convolutional Neural Networks as a means of identifying the leaf in the image space will be examined. To demonstrate the effectiveness of MDA as a control scheme, the typical elements of classical IBVS approach as well as MDA will be discussed for the purpose of comparison. This approach to visual servoing was tested using an experimental setup and the results will be discussed.

4.1 Leaf Detection

Object detection—a fundamental problem in Computer Vision—consists of producing the bounding boxes around the objects of desired class on an image. Traditional Machine Vision techniques are generally not suitable for the detection of objects like leaves. First, the leaves on the plants have complex backgrounds that often include other leaves, which makes it hard to separate an individual leaf from the background. Second, leaves have complex shapes and appearances, thus creating a model and estimating the parameters is not computationally feasible for a real-time application. Finally, since the plants are located in the field there is a wide variation in natural illumination. That excludes many color-based approaches that have to be finely tuned to the lighting conditions. Hence, the problem of leaf detection requires use of Machine Learning algorithms. In recent years Convolutional Neural Networks have become a popular approach.

To train the model we accumulated a small dataset of leaves images cropped to 227 × 227, labeled with two classes (leaves and background). Our network is based on the AlexNet architecture (Krizhevsky et al., 2012) and consists of five convolutional layers and two fully-connected layers. The model was fine-tuned by initializing with weights trained on a large ILSVRC12 dataset (1.2 million images), and then trained on our task specific dataset of leaves. This common technique greatly improves the robustness and accuracy of the model. During deployment the algorithm treats fully-connected layers as convolutional and, thus, gets the probability map for the target classes at once, instead of making many classification calls in a traditional sliding window approach.

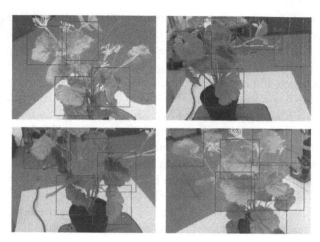

Fig. 10. Examples of leaves detections. The number in the corner of a bounding box indicates the score of the detected leaf.

In our particular problem, we do not have to detect all the leaves in the frame, but rather identify a good candidate for subsequent sampling, i.e., the one in front of the camera, not occluded, and not having extreme angles that are hard for the robotic arm to approach. That filtering is incorporated in the dataset creation (leaves at extreme angles are excluded from the dataset) and in selecting a candidate leaf processing (where after estimating 3D coordinates it can be dropped if deemed unsuitable). Figure 10 shows examples of leaf detections.

4.2 Leaf Tracking

To allow for successful execution of the leaf picking, there is a need to continuously track the target leaf from frame to frame as we move the robotic arm. Furthermore, triangulation of 3D coordinates requires precisely matched points within the target leaf for two consecutive frames. Since, during the execution, we have only one target leaf from the old frame and few candidate leaves in the new frame, we can efficiently obtain the SURF descriptors for these leaves (which would otherwise be computationally prohibitive). Then, to rule out the keypoints in the background we use green color thresholding on the leaf image; we also exclude the points in the corners, because they could belong to other leaves appearing within the bounding box.

For each pair of the old target leaf and a new candidate leaf, we perform quick matching with FLANN. We filter out the outliers with RANSAC algorithm that estimates the largest set of points that agrees with a perspective transformation. At the end of this procedure, the correct leaf

Fig. 11. Leaf tracking on several frames. Yellow bounding box indicates the target leaf that is being tracked; blue boxes indicate other detections in the frame; keypoints are highlighted on the leaves.

has significantly higher number of matches and is updated as our target. If all the leaves have low number of matches, we assume that the target leaf is missing in the current frame. We continue searching for the leaf for an arbitrary number of frames, and if it is not found, we pick a new target. Refer to Fig. 11.

This approach, typically, produces over one hundred matched feature points within a leaf. The points, however, are consistent only between one pair of consecutive frames, which makes it not suitable for IBVS. Identifying consistent features on a leaf is difficult because for each frame a given leaf will change its position and orientation within the bounding box, and on some occasions parts of the leaf can be outside of the bounding box. The usage of bounding box regression (Kuo et al., 2015) and semantic segmentation-aware models (Gidaris and Komodakis, 2015) will provide more precise localization of the leaf and may allow for identification of the four corners of the leaf along the axes. The performance of the aforementioned approaches on leaves and the extent to which it may help to draw up a constant feature detection method will be evaluated in our future work.

4.3 Visual Servoing

Different forms of visual-servoing can be used which would allow a manipulator to interact with an unstructured object. However, each form has varying strengths and weaknesses which would affect the performance of the robotic arm. In this analysis, two forms of visual-servoing were considered for implementation: Classical Image Based Visual Servoing and Monoscopic Depth Analysis (MDA). Some of the important factors in visual-servoing are formulation of the error vector, analysis of feature points, and control of the manipulator. These visual servoing methods will be introduced and then compared in the context of grasping a leaf from a plant.

Image Based Visual Servoing (IBVS) is a wide field of analysis that attempts to convert a Cartesian Space application into the image space. Numerous modifications of this method can be implemented to better fit a control scheme to achieve the desired task. The main strength of IBVS is that camera calibration is not necessary, nor does it require a prior information about the object being viewed (Chaumette and Hutchinson, 2006). This method attempts to minimize the difference between the feature points of an image with a desired set of feature points, thus orienting the camera in some desirable fashion.

4.4 Monoscopic Depth Analysis

Using two images to determine a feature point's Cartesian position has been well established in the field of image processing. Indeed, this method forms the basis of stereo cameras (Kalghatgi, 2012). In stereo-vision, two cameras, that are a known distance and orientation from one another, each take an image simultaneously. The images are compared, and, with information about the properties of the cameras, the 3D Cartesian position of corresponding feature points are determined. Typically, cameras used in stereo-vision will be pointed in the same direction and only be separated by a fixed distance between their optical axes; however, it can easily be shown that any known translation and rotation between cameras can be factored to determine corresponding feature point's Cartesian position, relative to a camera (similar to the setup in Fig. 13).

4.5 Error Vector Formulation

In IBVS, the error vector exists in the image space. When visualizing an object, a set of feature points are identified and compared against a desired position, such as shown in Fig. 12. The relationship between the pixel velocities and Cartesian velocities can then be quantified as follows: $ds/dt = Jf\, d(p_c)/dt$, where ds/dt is the rate of change in pixels, J_f is the feature Jacobian, and p_c is the velocity Twist vector of the camera in R^6 space, concatenating the translational and rotational velocity vectors yields $p_c = [x\ y\ z\ \Theta_x\ \Theta_y\ \Theta_z]^T$.

Since the error vector is formulated in the image space, the geometric structure of the object need not be known. Also, the distance to the object is not a necessary condition for convergence, an estimated value can be used (Chaumette and Hutchinson, 2006); however, accurate estimates do help in convergence. The result is that IBVS is well suited for unstructured objects.

The error formulation process is quite different in MDA. In MDA, the error vector is the difference in 3D space between the end-effector location and a desired position. While in IBVS, the current position and the desired position were known in the image space relative to the target (leaf), MDA assumes that only the current position and rotation in Cartesian space is known. The desired position is determined through the process of

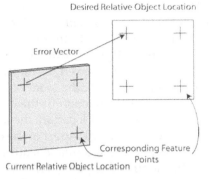

Fig. 12. An object being seen by a camera in both its current position and desired position. When the object is in the desired location relative to the camera, it will have a specific pattern of feature points.

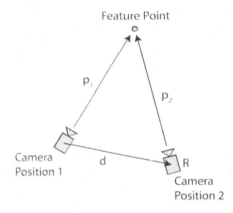

Fig. 13. Diagram depicting a stereo-vision camera set-up.

estimating the location of the feature points in Cartesian space. In this recursive process, the error vector is estimated after a pair of successive images are taken; simultaneously calculating a subsequent movement vector to minimize the error in the system. Figure 13 demonstrates the relevant variables in this triangulation process. Unlike IBVS, determining the exact depth of the object is critical in the process. However, like IBVS, the structure of the object does not need to be known ahead of time for convergence to be possible. MDA is also very applicable to unstructured visual-servoing applications.

4.6 Feature Points

Image Based Visual-Servoing and Monoscopic Depth Analysis both rely on distinguishable feature points determined by image processing methods. However, the constraints with which feature points can be used are widely different between the two methods. IBVS relies on a set of at least four distinguishable feature points that consistently exist in each frame for analysis. Furthermore, these feature points have to correspond to known aspects of the object being viewed. Examples of possible feature point in this application might include the stem of the leaf or the tip of the leaf. Since the error vector is a function of the desired feature points location and the known feature points location, the ability to correspond what is seen in the image with a known aspect of the object is essential.

In MDA, the constraints on the acceptable feature points are relaxed. In order to estimate the depth of a point, only a single feature point corresponded between two images is necessary. To estimate the location and rotation of an object, only three points are needed in two consecutive images. Furthermore, while having information about a feature point is always beneficial, MDA does not require the feature points to have a known relationship to the object ahead of time (beyond the assumption that the feature points belong to an object). Also, MDA does not require that the same feature points be identified in multiple successive images. MDA only requires that a set of feature points be corresponded in a pair of images; different sets of feature points can be used in each successive pair.

4.7 Robot Control

All visual-servoing techniques must address how the robotic arm is going to interpret information from the imaging routine. Both IBVS and MDA assume an eye-in-hand set-up of the camera and arm (indicating that the camera is rigidly attached to the end-effector of the robotic arm) and that the camera is the only sensor capable of detecting the object. However, just as the error vector domains differ between the two methods, so might the control domains. Controlling a system directed by IBVS allows for a certain level of flexibility. IBVS inherently lies in the image-space, but the control vector that is being manipulated is the Twist of the camera (and subsequently the end-effector). Thus, the commands of the control scheme can either be interpreted in the Cartesian space or the joint space of the arm. The ability to set the control domain allows for the best possible control of the arm to be chosen. Since the error vector of MDA exists in Cartesian space, the control of the arm must also exist in Cartesian space. This indicates that the arm control must be translated from joint space into Cartesian space, just as the feature points must be converted from image space into Cartesian space. While control can then be moved to a different domain, these initial conversions are necessary and impose a certain amount of limitations to the control.

4.8 Results

Object detection algorithm was able to detect at least one leaf for every frame where the plant was present. Once a leaf was picked as a target, that same leaf was consistently tracked in subsequent frames while being approached by the robotics arm. The performance of tracking algorithm was less robust and at times it failed to find matched points on the detected leaf in the new frame, typically due to substantial horizontal motion. This issue was accommodated by adjusting the robotic arm's search pattern.

The algorithms' computational performance was tested on a laptop with Intel i7 x 8 CPU and NVIDIA GTX880M GPU. The images from the camera had a resolution of 1280 × 720. Image Processing took on an average 0.7s per frame if run on GPU, with tracking using most of the time; and 7.0s on CPU, with the detection running the longest.

Using the Mico arm from Kinova, a six degree-of-freedom serial actuator with two fingers, and a camera fitted with a custom mount attached to the end-effector, a series of trials were performed to demonstrate the effectiveness of Monoscopic Depth Analysis. The visual-servoing approach can be divided into the following subtasks:

(1) Search for target leaf

(2) Approach target leaf

(3) Grasp target leaf

In this process, "Searching" for the target leaf is a process of exploring the area where the plant is believed to be until the image processing algorithm identifies a suitable candidate, which will be referred to as a target leaf. During the "Approach", the manipulator moves the camera to get various scene vantages in attempts to perform triangulation of desired feature points. If the target leaf is lost, or if correspondence cannot be performed, this step is canceled and the algorithm returns to the "Searching" stage. Finally, if the target leaf is identified, the manipulator attempts to grasp the leaf. The process of grasping the leaf is quite involved and beyond the scope of this experiment, as care has to be taken in how the manipulator grasps the leaf. If the approach does not take into consideration the orientation of the leaf or surrounding objects, it is possible to either miss the leaf entirely or accidentally push the leaf out of the way. For this experiment, the grasping stage is approximated by having the manipulator position its end effector at the location of the leaf without necessarily attempting to interact with the object.

In this experiment, it was considered a success if the manipulator was positioned less than a quarter inch away from the body of the leaf after the "Grasping" stage. A small number of trials were performed and are listed in Table 1. The trials were performed on a plastic, surrogate

Table 1. Grasping performance.

Searches Performed: 10
Successful Searches: 10
Approaches Performed: 23
Successful Approaches: 16
Average failure distance: 2 in (approx.)

plant in the following manner. Ten "Searches" were conducted; during each "Search", the robotic arm was positioned to view the plant from a variety of perspectives. If a leaf was identified and found to be acceptable (in terms of estimated distance to the arm), the arm would enter the "Approach" stage and attempt to accurately locate the leaf. If the leaf could not be identified in a subsequent frame, or if the algorithm decided that the error vector was not converging, the "Approach" stage would be abandoned and the algorithm would return to the "Searching" stage. If the error vector to the leaf converged, the "Approach" stage was advanced to the "Grasping" stage, an open-loop routine based on estimates of the position of the leaf. Although the leaf was sometimes "grasped" in the traditional understanding (the fingers of the manipulator pinched the leaf, as seen in Fig. 14), this was not the stated goal of this experiment. For this experiment, the "grasp" was considered a success if the fingers of the robotic manipulator touched the desired leaf. Following these steps, multiple "Approaching" stages were performed for each "Searching" stage. Thus, even though each individual attempt at approach did not always lead to a success, each "Searching" stage did end with a leaf being "grasped" at least once. The average failure distance was difficult to measure; the value recorded is meant to give an impression as to the magnitude of failure rather than a precise average distance. Although the control algorithm for grasping the leaf can be refined, these initial results show the promise of the MDA technique for finding leaves in 3D space. The manipulator was able to demonstrate an ability to locate and interact with unstructured objects with consistency.

As we move forward, there is room for further improvement in perception and control algorithms. First, more robust detections can be achieved by using current state-of-the-art object detection approaches (e.g., Faster-RCNN), which use region proposals to classify the image patches that are likely to have objects. Then, the feature points required by IBVS could be approximated by the corners of the bounding box produced by object detection, thus enabling combining both control approaches previously discussed. In particular, IBVS can be used to quickly start approaching the leaf during first few iterations, while simultaneously

Fig. 14. Example of a successful leaf grasp by the robot arm. The leaf was located visually, its position estimated, and then the arm's end-effector was commanded to the leaf position.

building the distance estimate; then switching to MDA will allow for accurate positioning and grasping.

4.9 Dual-Arm Visual Servoing for Fruit Picking

To conclude our discussion of visual servoing robot control, we include a brief description of a related agricultural application (apple picking) that, similar to field scouting, takes place in an unstructured environment. The shape of the branches of any given apple tree (see Fig. 15) is unique and difficult to characterize geometrically. In the context of apple picking, scattered fruit targets of interest are interspersed among the tree's branches and the goal is to successfully reach the targets to grasp them.

We approach this problem by combining two serial manipulators, each equipped with an eye-in-hand camera, which examine the tree architecture to find a clear path to the fruit. The two manipulators are designated as the "Search Arm" and the "Grab Arm" (Fig. 16). The Search Arm is assumed further from the tree and clear of obstacles, while the Grab Arm is set among the branches where obstacles are present. The Search Arm provides

Fig. 15. Simulation of apple picking using dual, coordinated robot arms controlled by eye-in-hand visual servoing.

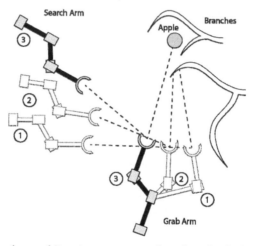

Fig. 16. Illustration of several iterative movements that allow the Grab Arm to explore the unstructured tree branch environment.

a wider view of the scene and the grab arm has a narrower view, focusing on the fruit. The two arms work together to navigate the branches of the tree and to bring the Grab Arm to the fruit. The system operates by using machine vision to look for unobstructed views of the Grab Arm and the apples. By compiling the clear volumes within the tree branches, the arms are able to find a navigable path to the fruit relatively quickly. The tree is examined in sections, and the grab arm picks as many apples as possible within its current section. Simulations demonstrate that with two arms working in tandem, each with a distinct but cooperative task, the arms are able to identify viable paths to pick apples set among the branches of a tree. Field testing will be underway soon.

5. CONCLUSIONS

Agricultural robotic systems need to operate effectively in unstructured environments that vary both spatially and temporally. We have presented a case study of an automated "field scout" ground platform equipped with the means for both sensing and manipulating its changing environment for the purpose of providing actionable data to a farmer. The technical topics we have covered include: (1) 4-dimensional mapping using 2-dimensional imaging and (2) robot arm end-effector manipulation using visual servoing control for: (a) leaf picking in peanut plant rows and (b) apple picking using two coordinated robot arms. Results from both simulations and field experiments were described and evaluated, showing successful outcomes that will serve as the foundation for future work in this emerging field.

REFERENCES

Agarwal, S., Snavely, N., Simon, I., Seitz, S. M., and Szeliski, R. 2009. Building Rome in a day. International Conference on Computer Vision (ICCV). pp. 72–79.

Badino, H., Huber, D., and Kanade, T. 2011. The CMU visual localization data set. http://3dvis.ri.cmu.edu/data-sets/localization.

Bergtold, J. S., Raper, R. L., and Schwab, E. B. 2009. The economic benefit of improving the proximity of tillage and planting operations in cotton production with automatic steering. Applied Engineering in Agriculture. 25(2): pp. 133–143.

Carlevaris-Bianco, N., Ushani, A. K., and Eustice, R. M. 2016. University of Michigan north campus long-term vision and lidar dataset. Int. J. of Robotics Research. 35(9): pp. 1023–1035.

Chaumette, F., and Hutchinson, S. 2006. Visual Servoing Part I: Basic Approaches. IEEE Robotics and Automation Magazine. 4: pp. 82–90.

Fallon, M. F., Johannsson, H., Kaess, M., Rosen, D. M., Muggler, E., and Leonard, J. J. 2012. Mapping the MIT stata center: Large-scale integrated visual and RGB-D SLAM. In RSS Workshop on RGB-D: Advanced Reasoning with Depth Cameras.

Font, D., Tresanchez, M., Martınez, D., Moreno, J., Clotet, E., and Palacın, J. 2015. Vineyard yield estimation based on the analysis of high resolution images obtained with artificial illumination at night. Sensors 15(4): pp. 8284–8301.

Furukawa, Y. and Ponce, J. 2010. Accurate, dense, and robust multi-view stereopsis. IEEE Trans. Pattern Anal. Machine Intell. 32(8): pp. 1362–1376.

Gidaris, S., and Komodakis, N. 2015. Object detection via a multi-region & semantic segmentation-aware CNN model. http://arxiv.org/abs/1505.01749.

Hague, T., Tillett, N., and Wheeler, H. 2006. Automated crop and weed monitoring in widely spaced cereals. Precision Agriculture. 7(1): pp. 21–32.

Hartley, R. I., and Zisserman, A. 2004. Multiple View Geometry in Computer Vision. Cambridge University Press.

Indelman, V., Nelson, E., Dong, J., Michael, N., and Dellaert, F. 2016. Incremental distributed inference from arbitrary poses and unknown data association: Using collaborating robots to establish a common reference. IEEE Control Systems. 36(2): pp. 41–74.

Kalghatgi, R. 2012. Reconstruction techniques for fixed 3-D lines and fixed 3-D points using the relative pose of one or two cameras. Thesis, Georgia Institute of Technology.

Krizhevsky, A., Sutskever, I., and Hinton, G. E. 2012. Imagenet classification with deep convolutional neural networks. *In*: Pereira, F., Burges, C. J. C., Bottou, L., and Weinberger, K. Q. (eds.). Advances in Neural Information Processing Systems. 25: pp. 1097–1105.

Kuo, W., Hariharan, B., and Malik, J. 2015. Deepbox: Learning objectness with convolutional networks. pp. 2479–2487. *In*: Proceedings of the IEEE International Conference on Computer Vision.

Lowe, D. 2004. Distinctive image features from scale-invariant keypoints. Int. J. of Computer Vision. 60(2): pp. 91–110.

Muja, M., and Lowe, D. G. 2014. Scalable nearest neighbor algorithms for high dimensional data. IEEE Trans. Pattern Anal. Machine Intell. 36(11): pp. 2227–2240.

Nuske, S., Wilshusen, K., Achar, S., Yoder, L., Narasimhan, S., and Singh, S. 2014. Automated visual yield estimation in vineyards. J. of Field Robotics. 31(5): pp. 837–860.

Rains, G. C., Faircloth, A. G., Thai, C., and Raper, R. L. 2014. Evaluation of a simple pure pursuit path-following algorithm for an autonomous, articulated-steer vehicle. Applied Engineering in Agriculture. 30(3): pp. 367–374.

Sa, I., Ge, Z., Dayoub, F., Upcroft, B., Perez, T., and McCool, C. 2016. Deepfruits: A fruit detection system using deep neural networks. Sensors. 16(8): p. 1222.

Subramanian, V., Burks, T. F., and Dixon, W. E. 2009. Sensor fusion using fuzzy logic enhanced Kalman filter for autonomous vehicle guidance in citrus groves. Trans. of the ASABE. 52(5): pp. 1411–1422.

Vellidis, G., Ortiz, B., Beasley, J., Hill, R., Henry, H., and Brannen, H. 2013. Using RTK-based GPS guidance for planting and inverting peanuts. pp. 357–364. *In*: J.V. Stafford (ed.). Precision Agriculture 2013—Proceedings of the 9th European Conference on Precision Agriculture.

4

Current and Future Applications of Cost-Effective Smart Cameras in Agriculture

Young K. Chang[†,*] and *Tanzeel U. Rehman*[†]

1. INTRODUCTION

Relentless population increase will result in over 9 billion predicted population on the globe by 2050 (FAO, 2015). However, yield gain of major cereal crops, even with the help of a mechanization, farm enlargement and/or technology has plateaued in last two decades (Grassini et al., 2013). Eighty four percent of the world farms are less than 2 hectares and many of these farms lack the financial ability to adapt emerging technologies (like, multispectral camera, hyperspectral camera, etc.) to increase their productivity (FAO, 2014). Another alarming situation facing the agricultural industry is that the total number of farm operators in the world are constantly declining and we are facing agricultural work force aging phenomenon (ILO, 2014).

A smart camera is an intelligent vision system that not only acquires images but also extracts useful information, applies algorithms and makes decisions for specific applications including automation (Belbachir and Göbel, 2009). Use of smart camera started in 1990s by many industrial sectors and it has also been studied for agricultural purposes, because it is non-destructive, rapid, efficient and cost-effective which will decrease human labor. The smart camera can be used for saving agro-chemicals,

Engineering Department, Faculty of Agriculture, Dalhousie University, Truro, Nova Scotia, Canada.
Email: Tanzeel.Rehman@dal.ca
[†] These authors contributed equally to this chapter.
[*] Corresponding author: YoungChang@dal.ca

non-invasive means of performing a particular task, minimizing the cost and labor associated with the sorting, etc. However, most of the studies on smart camera or machine vision in agriculture are focused on expensive hyperspectral or multispectral cameras, which are too expensive for most of small farmers. Therefore, current use and potential future use of cost-effective smart camera in agriculture was reviewed in this chapter.

The applications of cost-effective smart cameras for sorting of fruits, vegetables and grains by using color, shape and textural features are introduced in Section 3. Subsequently, hardware based image processing tools for minimizing the computational requirements are presented in Section 4. The Section 4 shows the potential of digital signal processor (DSP), field programmable gate array (FPGA), advanced reduced instruction set computing machine (ARM) and graphic processing unit (GPU) for future agricultural applications. This chapter covers only the above mentioned applications of smart camera, other applications like yield monitoring, plant phenotyping, disease detection, etc. are not reviewed.

2. SMART CAMERAS FOR WEED—CROP SEGMENTATION

Weeds are one of the major yield limiting factors in almost all the cultivated and non-cultivated crops around the world. They can holdup the plant nutrients, compete with plants, harbor diseases and insects, and may hinder the harvesting operation (Kinsman, 1993). The most common and commercially available weed management protocols include the blanket spraying of herbicides across the entire field, thereby raising the environmental concerns (Kazmi et al., 2015) and cost of production (Meyer, 2011). The variability in the spatial distribution of weeds can be detected in non-invasive manner using the state-of-the-art cutting-edge sensing tools available commercially (Andújar et al., 2013). The 'sensed' information can be used to trigger the robust 'controllers' to target the individual weed canopy thereby, providing the means to minimize the cost of production.

Weed locations and spatial variability within fields can either be coarsely 'sensed' using remote sensing platforms or 'fine sensing' procedures' based on near-ground methods can be opted for real-time applications (Pérez et al., 2000). The remotely sensed aerial spectral scans can be used to develop the prescription maps in geo-statistical software to vary the application rates of pesticides according to the weed location (Michaud et al., 2008). However, these systems mainly rely on the good quality, up-to-date aerial data followed by the comprehensive data management and processing for weed spots detection (Chang et al., 2012)

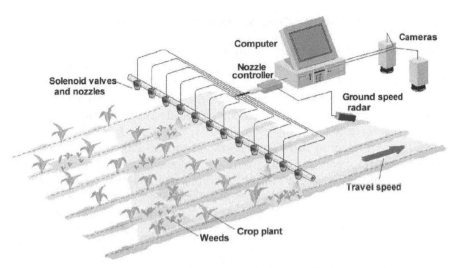

Fig. 1. The Smart sprayer concept. The system includes a multiple-camera vision system, a ground speed sensor and a nozzle controller (Reprinted from Computers and Electronics in Agriculture. 36(2), Tian, Development of a sensor-based precision herbicide application system, 133–149. Copyright (2002), with permission from Elsevier).

thereby not suitable for the real-time field applications. Alternatively, optoelectronic sensors attached to ground vehicles can be used to develop the spectral signatures by using time of flight approach for the reflected optical beams from different crop and weed areas (Andújar et al., 2011). These sensors, however, were not able to discriminate between the crop and weeds of the same height and thus could only be used to discriminate the vegetative area from the bare soil (Andújar et al., 2013). Therefore, many researchers have used smart cameras for the segmentation of weeds from crops (Shearer and Holmes, 1990; Woebbecke et al., 1995a; Meyer et al., 1998; Burks et al. 2000; Lamm et al., 2002; Meyer and Neto, 2008; Ahmed et al., 2011; Guerrero et al., 2012; Kazmi et al., 2015). Figure 1 shows the concept of the smart sprayer utilizing segmentation of weeds from crops (Tian, 2002). These cameras can be used with variety of image processing algorithms to exploit the color, shape and textural information contained in the acquired images. These algorithms can be used to explore the different traits related to plant and weed canopies along with bare soil information. Furthermore, unlike remotely sensed data, these methods are not influenced by the positional error and don't require any pre-processing and prescription maps development. The following sub-sections explain the use of different image processing algorithms for sensing the spatial variability in the weed location and their advantages and drawbacks for real-time applications.

2.1 Color Based Weed—Crop Segmentation Systems

Analysis of the different vegetation by exploiting its color related spectral attributes is perhaps one of the easiest way to discriminate plants from background clutter. Color spectral information contained in an agronomic image can be utilized to discriminate the vegetative biomass and soil residues (Woebbecke et al., 1995a; Lamm et al., 2002; Hague et al., 2006; Meyer and Neto, 2008; Guijarro et al., 2011; Guerrero et al., 2012; Kazmi et al., 2015). An array of different vegetation indices has been evolved and tested over the years by using available information in different visible spectral channels and their combinations for weed-crop discrimination (Woebbecke et al., 1995a; El-Faki et al., 2000; Mao et al., 2003; Kataoka et al., 2003; Meyer, 2011; Montalvo et al., 2013; Chang et al., 2014; Esau et al., 2014). Whilst most of these indices amplify the information contained in a respective color channel (Fig. 2); thereby accentuating the color of any particular region of interest (Meyer and Neto, 2008; Meyer, 2011).

Earlier applications of vegetation indices involved the non-normalized red-green-blue (RGB) color coordinates. These coordinates were largely influenced by the camera parameters and amount of incident illumination

Fig. 2. Comparison of vegetative indices (ExGExR, ExG, and NDI) and hand extracted mask (Reprinted from Computers and Electronics in Agriculture. 63(2), Meyer and Neto, Verification of color vegetation indices for automated crop imaging applications, 282–293. Copyright (2008), with permission from Elsevier).

on the terrain and were not very effective in discriminating green biomass from background residues (Woebbecke et al., 1995a). Woebbecke et al. (1995a) also normalized RGB chromatic coordinates by applying the variation in intensities uniformly across all three color channels (Cheng et al., 2001). The normalized color coordinates were successfully used to develop and compare the performance of different vegetation indices with the excess green (ExG) as an optimal selection for separating the plants from residues (Woebbecke et al., 1995a; Lamm et al., 2002; Mao et al., 2003; Guerrero et al., 2012; Kazmi et al., 2015).

Meyer et al. (1998) adopted an alternative approach for identifying the soil and residue by amplifying the redness color image thus resulting in an excess red (ExR) index. An improved color based index (ExGExR, ExG-ExR or ExGR) was achieved by taking the difference of ExG and ExR (Camargo, 2004; Meyer and Neto, 2008). This index generates binary images without using any manually defined threshold level and performed comparably to the ExG index (Meyer and Neto, 2008). Marchant et al. (2001) used a red/green ratio to detect vegetation against soil background and it was compared with red/Near Infra-Red (NIR) ratio and a new classification method (alpha-method). A color index of vegetation extraction (CIVE) was used to estimate the growth of soybean and sugar beet and was found to have a high degree of correlation with manually measured plant parameters (Kataoka et al., 2003). The same color index was reported in number of studies to identify the pixels containing background soil and crop residue from the plant pixels (Guijarro et al., 2011; Guerrero et al., 2012; Montalvo et al., 2013; Kazmi et al., 2015; Yang et al., 2015). The combination of this color index with ExR, ExGExR, NDI, GB, RBI, ERI, EGI, Rn and Gn resulted in a high classification accuracy of 97.83% (Kazmi et al., 2015). Hague et al. (2006) came up with a more intense and color illumination resistant vegetative index (VEG) by studying the physics of image formation with respect to cereal fields.

The weighted average of four existing vegetation indices (ExG, ExGExR, CIVE and VEG) were combined to analyse the information regarding the greenness of agricultural field images (Guijarro et al., 2011). The results of these individual color indices for identifying the green weeds indicated that CIVE achieved the highest identification accuracy and was therefore given highest weight when combining these four color indices. A similar combination of ExG, CIVE and VEG was proposed as a solution to extract the green plant pixels masked by the red spectral component of soil (Guerrero et al., 2012). The binary image of green plant pixels was generated using Otsu's thresholding algorithm (Otsu, 1979) followed by the support vector machine (SVM) based classification technique. An automated expert system (AES) used same color index based algorithm to delineate the green plants from the background (Montalvo et al., 2013).

The results of their study showed that AES outperformed an iterative approach of applying two different thresholds to differentiate between different classes (Demirkaya et al., 2008) and the SVM (Guerrero et al., 2012). Burgos-Artizzu et al. (2011) modified the linear combination of coefficients of RGB color planes to improve the performance of the ExG. The resultant gray scale image was binarized by using the histogram mean intensity threshold followed by morphological opening and area threshold to identify inter-row weeds in maize fields. The results showed that this system was able to correctly identify 85.1% of the inter-row weeds and 68.9% of the maize crop rows.

Golzarian and Frick (2011) developed four indices by combining different color characteristics of the RGB images for laboratory evaluation, however, none of these performed appropriately during the field application (Kazmi et al., 2015). Chang et al. (2014) reported green ratio (G-ratio) index to identify newly emerging green weeds (grasses) against soils and reddish pruned wild blueberry plants. Esau et al. (2014) used the same index to apply the fungicides on wild blueberry plants in real-time. Kazmi et al. (2015) used 14 different color indices for the identification of the creeping thistle in sugar beet field images. The ability of the individual index was compared with others followed by the combinations of different color indices for the weed identification. The results of this study indicated that linear discriminant analysis (LDA) with stepwise regression was able to correctly identify 97.83% of thistles.

In addition to the indices from RGB color space, other color spaces (Woebbecke et al., 1995a; Tang et al., 2000; Golzarian et al., 2012; Bai et al., 2013; Kim et al., 2015; Yang et al., 2015) and non-visible bands of electromagnetic spectrum (Haggar, 1983; Guyer et al., 1986; Shearer and Jones, 1991; Franz et al., 1991; Gerhards and Oebel, 2006; Hunt et al., 2011) were used to locate the position of the weeds within crop rows. Woebbecke et al. (1995a) used modified hue (MH) component of a RGB image to distinguish plants from its background and to identify the monocot from the dicot. The results indicated that the MH was able to delineate the plants from background but the monocot identification from the dicot was not successful. Tang et al. (2000) implemented a Hue, Saturation and Intensity (HSI) color space based genetic algorithm capable of finding the green plants by searching for the global optima.

A real-time crop-residue segmentation algorithm was developed by analyzing the illumination effected field images under RGB, HSI, $I_1I_2I_3$, YC_bC_r and CIE Lab color space along with the iterative threshold determination technique (Ji et al., 2007). The best segmentation of plant images was achieved by the H, a, I3 and Cr components of these color models for simple backgrounds. Two variants of environmentally adaptive segmentation algorithm (EASA) were developed using the HSI

Table 1. Comparison of different color channels and vegetation indices based weed-crop segmentation systems.

Color Index	Mathematical Relationship	Crop	Segmentation Technique	Reference
NDI	$(G - R)/(G + R)$	Cereal crops, sugar beet	Manually selected local threshold with morphological dilation	Woebbecke et al., 1993; Pérez et al., 2000; Meyer and Neto, 2008; Kazmi et al., 2015; Hamuda et al., 2016
R_n, G_n, B_n	$R_n = R/(R + G + B)$ $G_n = G/(R + G + B)$ $B_n = B/(R + G + B)$	Sugar beet	Otsu's threshold with linear discriminant analysis and Mahalanobis distance	Woebbecke et al., 1995a; El-Faki et al., 2000; Kazmi et al., 2015
ExG	$2G_n - R_n - B_n$	Wheat, corn, sugar beet	Otsu's threshold, area threshold	Woebbecke et al., 1995a; Camargo, 2004; Nieuwenhuizen et al., 2007; Meyer and Neto, 2008; Guijarro et al., 2011; Kazmi et al., 2015; Hamuda et al., 2016
GB	$G_n - B_n$	Wheat, corn	Otsu's threshold, area threshold	Woebbecke et al., 1995a
RG	$R_n - G_n$	Wheat, corn	Otsu's threshold, area threshold	Woebbecke et al., 1995a
GB/RG	$(G_n - B_n)/|R_n - G_n|$	Wheat, corn	Otsu's threshold, area threshold	Woebbecke et al., 1995a
ExR	$1.4R_n - G_n$	Soybean, sunflower	Otsu's threshold, area threshold	Meyer et al., 1998; Camargo, 2004; Kazmi et al., 2015; Hamuda et al., 2016
RG ratio	R/G	Vegetation		Marchant et al., 2001

Table 1 cont. ...

...Table 1 cont.

Color Index	Mathematical Relationship	Crop	Segmentation Technique	Reference
CIVE	$0.441R_n - 0.811G_n + 0.385B_n + 18.78745$	Soybean, sugar beet	Threshold with discriminant analysis	Kataoka et al., 2003; Guijarro et al., 2011; Kazmi et al., 2015; Yang et al., 2015; Hamuda et al., 2016
ExGExR (ExGR)	$(2G_n - R_n - B_n) - (1.4R_n - G_n)$	Soybean, sunflower	Self-generating automatic threshold followed by Gaussian filter	Camargo, 2004; Meyer and Neto, 2008; Guijarro et al., 2011; Kazmi et al., 2015; Hamuda et al., 2016
Gray	$0.2989R_n + 0.5870G_n + 0.1140B_n$	Grasslands, sugar beet, wild blueberry	Area threshold/Otsu's threshold with linear discriminant analysis	Gebhardt et al., 2006; Gebhardt and Kühbauch, 2007; Chang et al., 2012; Kazmi et al., 2015
VEG	$G_n / (R_n^{0.667} - B_n^{0.333})$	Winter wheat	Manually selected global threshold	Hague et al., 2006; Guijarro et al., 2011; Kazmi et al., 2015; Yang et al., 2015; Hamuda et al., 2016
RB	$R - B$	Sugar beet	Combination of K-means clustering and a Bayes classifier, adaptive Resonance Theory 2 (ART2) Neural Network for Euclidean distance-based clustering	Nieuwenhuizen et al., 2007
COM1	$0.25ExG + 0.3ExGExR + 0.33CIVE + 0.12VEG$	Barley, corn	Mean histogram threshold	Guijarro et al., 2011; Yang et al., 2015; Hamuda et al., 2016

MExG	$1.26G_n - 0.884R_n - 0.311B_n$	Maize	Otsu's threshold, Mean histogram threshold with morphological opening and area threshold	Burgos-Artizzu et al., 2011; Hamuda et al., 2016
COM2	$0.36ExG + 0.47CIVE + 0.17VEG$	Maize	Otsu's threshold with support vector machines	Guerrero et al., 2012; Montalvo et al., 2013; Hamuda et al., 2016
RBI	$(R_n - B_n)/(R_n + B_n)$	Wheat, sugar beet	Global threshold from hue image followed by principal component analysis	Golzarian and Frick, 2011; Kazmi et al., 2015
ERI	$(R_n - G_n) \times (R_n - B_n)$	Wheat, sugar beet	Global threshold from hue image followed by principal component analysis	Golzarian and Frick, 2011; Kazmi et al., 2015
EGI	$(G_n - R_n) \times (G_n - B_n)$	Wheat, sugar beet	Global threshold from hue image followed by principal component analysis	Golzarian and Frick, 2011; Kazmi et al., 2015
EBI	$(B_n - G_n) \times (B_n - R_n)$	Wheat, sugar beet	Global threshold from hue image followed by principal component analysis	Golzarian and Frick, 2011; Kazmi et al., 2015
G-ratio	$255G/(R + G + B)$	Wild blueberry	Manually selected global threshold	Chang et al., 2014; Esau et al., 2014

color space for handling the sunflower images taken under complex field conditions (Ruiz-Ruiz et al., 2009). The performance of these variants in terms of the segmentation efficiency were not significant. A mean shift algorithm with back propagation neural network (BPNN) was applied on the color features extracted from the RGB and HSI color planes to classify between plant and non-plant regions (Zheng et al., 2009). A median of miss-segmentation was about 4.2%.

A crop color model was developed in the CIE Lab color space for segmenting the rice crop images under complex illumination conditions (Bai et al., 2013). The mathematical morphology based learning technique was used to relate the color of the green plants to its mean pixel lightness (L) for developing a weed segmentation criterion. The comparison of this approach with other RGB based indices indicated the superiority of the proposed approach with mean segmentation rate of 87%. A similar approach, based on the combination of ExR and CIE Lab color space was used to segment the soybean plant pixels from residual pixels (Kim et al., 2015). The binary images were developed by using the Otsu's and triangle threshold methods. The algorithm was capable of achieving a segmentation accuracy of 98%.

2.2 Shape Based Weed—Crop Segmentation Systems

Identification of weed species and estimation of their densities on the basis of difference in spectral reflectivity may not yield the desired results due to the similar reflective signatures of crop and weed especially during the initial growing season (Andreasen et al., 1997). Another reason for relatively less reliable classification accuracies was found to be variation in luminance and color temperature caused by the varying outdoor illumination conditions (Tian and Slaughter, 1998). The presence of transmitted light and inter-reflections along with the day light illumination conditions may also decrease the ability of color based segmentation approaches (Hague et al., 2006). Therefore, a process of identifying the weeds on the basis of shape of individual leaf or plant canopy was studied in greater depth (Kincaid and Schneider, 1983; Guyer et al., 1986; Franz et al., 1991, 1995; Woebbecke et al., 1995b; Chaisattapagon and Zhang, 1995; Lee et al., 1999; Pérez et al., 2000; Tian et al., 2000; Mathanker et al., 2007; Golzarian and Frick, 2011; Shinde and Shukla, 2014).

The geometrical orientation of individual plant leaf or canopy can be expressed by quantifying its shape descriptors such as length or width (Chi et al., 2003) and can ultimately lead to real-time pattern recognition and decision making. The supervised leaf shape signatures were developed by studying the leaf complexity and dissection index of reconstructed leaves using the normalized Fourier coefficients (Kincaid and Schneider, 1983).

Table 2. Transformed color spaces based vegetation indices for weed-crop segmentation.

Color Space	Crop	Segmentation Technique	Average Segmentation Accuracy	Reference
H	Wheat, corn	Otsu's threshold	Not reported	Woebbecke et al., 1995a
HSI	Soybean	Genetic segmentation algorithm	Not reported	Tang et al., 2000
RGB, HSI, $I_1I_2I_3$, YC_bC_r, CIE Lab	corn	Double peak threshold, Otsu's threshold, Iterative threshold	Not reported	Ji et al., 2007
HS and H	Sunflower	Environmentally adaptive segmentation algorithm	54.37% for EASA HS, 52.58% for EASA H	Ruiz-Ruiz et al., 2009
RGB and HSI	Soybean	Mean shift algorithm with back propagation neural network	96.8% with reduced segmentation rate on green plant parts with shadows	Zheng et al., 2009
LUV	Soybean	Mean shift algorithm with fisher linear discriminant analysis	97%	Zheng et al., 2010
H	Wheat	Manually selected global threshold	83.9%	Golzarian et al., 2012
HI	Maize	Affinity propagation hue and intensity look up tables	96.68%	Yu et al., 2013
CIE Lab	Rice	Morphological dilation and erosion based modeling	87.2%	Bai et al., 2013
CIE Lab	Rice, cotton	Particle swarm optimization clustering and morphological modeling	88.1% and 91.7%	Bai et al., 2013
CIE Lab	Soybean	Otsu's and tringle threshold	98%	Kim et al., 2015
HSV	Maize	HSV based decision tree	Reported as Good	Yang et al., 2015

Franz et al. (1991) studied the effect of leaf occlusion on the performance of a shape based weed detection system. The non-occluded leaves of soybean and different weeds were identified by aligning the leaf boundary curvature extracted at two leaf stage with respective curvature models. This study used Fourier-Mellin correlation to calculate resampling curvature functions for partially occluded leaves followed by their match with curvature models.

Guyer et al. (1993) considered the leaf and overall plant canopy shapes and achieved 69% correct identification rate for 40 weeds and soybean crop. The authors reported that no single shape feature alone was sufficient to distinguish different plant species. Woebbecke et al. (1995b) found similar results and reported that any particular shape feature did not work efficiently as a plant classifier, because of greater phonological variance among plants of the same species. Two grass species were segmented from the narrow-leaf wheat using a combination of color, shape and texture images followed by the principal component analysis (PCA). Amongst the variables selected for developing the PCA model by using a correlation matrix indicated that the color features contributed more towards the identification of weeds (Golzarian and Frick, 2011). Pérez et al. (2000) used a set of five geometrical features and seven normalized Hu moments (Hu, 1962) along with Bayesian and K-nearest neighbor classification rules. Both of them classified the crop successfully rather than weeds.

Identification of leaves on the basis of their shape descriptors showed high (> 90%) classification accuracies (Guyer et al., 1986; Woebbecke et al., 1995b; VijayaLakshmi and Mohan, 2016). However, using this methodology in the real-field condition is not possible because of the commingled leaf and canopy structures. The simple shape parameters along with kernel based Particle Swarm Optimization and Fuzzy relevance vector machines were reported by VijayaLakshmi and Mohan (2016) with an accuracy of 99.87%, but its adaptability and suitability for the realistic field applications still needs to be analyzed in greater depth. Three studies reported the real-time application of their algorithm in the field. The real-time applications of shape based descriptors for weed identification showed comparatively poor (< 75%) performance (Lee et al., 1999; Tian et al., 2000; Gebhardt et al., 2006). The major bottleneck in achieving the high classification accuracies appears to be the classical Bayesian classification rule as compared to neural network and SVM counterparts (Lin, 2009; Li and Chen, 2010; Herrera et al., 2014).

The newly emerging feature selection/optimization procedures along with the modern classification techniques showed more potential for dealing with the field scale variability in real-time manner (Lin, 2009; Li and Chen, 2010; Herrera et al., 2014). These approaches have also shown their potential for dealing with the complexity associated with partially

Table 3. Comparison of classical and modern shape based weed-crop segmentation techniques.

Selected Features	Crop	Classification Technique	Best Classifying Features	Maximum Classification Accuracy	Reference
(Perimeter)² / Area, Central Moment, Elongatedness, Principal axis moment of inertia, Minor axis moment of inertia	Soybean, tomatoes, corn	Bayes classification rule	Central Moment of individual leave	90.2% for laboratory scale analysis	Guyer et al., 1986
Roundness, Aspect ratio, Perimeter/Thickness ratio	Corn	Threshold defined by mean values of features	Aspect ratio, Perimeter/ Thickness ratio	Not Reported	Woebbecke et al., 1993
Roundness, Aspect ratio, Thickness, Elongatedness, Perimeter/Thickness ratio, ICM_1	Corn	Threshold defined by mean values of features	Aspect Ratio and ICM_1	74% and 89%	Woebbecke et al., 1995b
Eccentricity, Compactness, Eight Hu invariant moments ($\varphi_1, \varphi_2, \varphi_3, \varphi_4, \varphi_5, \varphi_6, \varphi_7, \varphi_8$)	Wheat	Multivariate discriminant Analysis	Eccentricity, Compactness, Three invariant moments ($\varphi_1, \varphi_2, \varphi_4$)	No: reported	Chaisattapagon and Zhang, 1995
Area, Major axis length, Minor axis length, centroid, Area/length, Compactness, Elongation, Perimeter/Broadness, Log_{10} (Height/Width), Sum of radius of curvature	Tomatoes	Bayesian Classification rule	Elongation and Compactness (Lee et al., 1999), Major axis length/ Perimeter, Elongation, Aspect ratio, Compactness (Tian et al., 2000)	73.-% for tomatoes and 68.8% for weeds (Lee et al., 1999) 65% (Tian et al., 2000)	Lee et al., 1999; Tian et al., 2000

Table 3 cont. ...

...Table 3 cont.

Selected Features	Crop	Classification Technique	Best Classifying Features	Maximum Classification Accuracy	Reference
Major axis length, Aspect ratio, Area, (Major axis length)^2/Area, Roundness, Seven Hu invariant moments ($\phi_1, \phi_2, \phi_3, \phi_4, \phi_5, \phi_6, \phi_7$)	Cereal	Bayesian rule and K-nearest neighbor	No feature selection procedure was carried out	74.5% and 79.2%	Pérez et al., 2000
Area, Area/Contour length, Length/Width, Circularity, Convexity, Maximum diameter, Roundness, Spikes	Cabbage, carrots	Fuzzy logic oriented model	Area, Area/Contour length, Circularity, Convexity, Maximum diameter, Roundness,	88% for Cabbage and 72% for Carrots	Hemming and Rath, 2001
Shape factor, Circularity, Eccentricity, Area, Perimeter	Grasslands	Maximum likely hood estimation	Shape factor, Circularity, Eccentricity	40% to 60%	Gbebhardt et al., 2006
Area, Perimeter, Equivalent diameter, Eccentricity, Shape Factor, Circularity	Wheat	Bayesian Classification rule	No feature selection procedure was carried out	40%	Mathanker et al., 2007
Area/Length, Compactness, Elongation, Aspect Ratio, Length/Perimeter, Log_{10} (Height/Width), Perimeter/Broadness	Corn	Minimum-Redundancy-Maximum-Relevance method with support vector machines	Area/Length, Perimeter/Broadness, Elongation, Aspect Ratio, Log_{10} (Height/Width),	94% for Corn and 99% for weeds	Lin, 2009
Compactness, Aspect ratio, Roundness, Elongation, Rectangularity, Seven Hu invariant moments ($\phi_1, \phi_2, \phi_3, \phi_4, \phi_5, \phi_6, \phi_7$)	Cotton	Ant colony optimization followed by support vector machines	Compactness, Aspect ratio, Elongation, ϕ_1 and ϕ_2	97.50%	Li and Chen, 2010

Features	Crop	Method	Selected features	Accuracy	Reference
Width, Waddle disk ratio	Wheat	Principal Component Analysis	Width and Waddle disk ratio	88% (Rye grass) and 85% (Brome grass)	Golzarian and Frick, 2011
Perimeter, Diameter, Minor axis length, Major axis length, Eccentricity, Area, Seven Hu invariant moments (φ_1, φ_2, φ_3, φ_4, φ_5, φ_6, φ_7)	Maize	Support vector machines, Choquet fuzzy integral, Sugeno fuzzy integral, Dempster-Shafer theory and fuzzy multi-criteria decision making	Major axis length, φ_1, φ_2 and φ_7	82.8%, 84.2%, 84.1%, 81.9% and 85.8% respectively	Herrera et al., 2014
Area, Centroid, Eccentricity, Equivalent diameter, Extent, Major axis length, Minor axis length, Orientation, Perimeter	ICL leaf dataset (Wang et al., 2015)	Kernel based Particle Swarm Optimization followed Fuzzy relevance vector machines	Not reported	99.87% in-combination with other color and texture features at laboratory scale	VijayaLakshmi and Mohan, 2016

occluded vegetative leaves. The cost associated with their computational complexities and overheads, however, needs to be handled properly in order to make them suitable candidates for real-time applications.

2.3 Texture Based Weed—Crop Segmentation Systems

Despite achieving very high accuracies on a single leaf, shape features were not able to perform properly on the leaf canopy because of the sever overlapping between inter- or intra-class leaves (Lee et al., 1999; Tian et al., 2000; Meyer, 2011). Moreover, the variability in outdoor illumination condition may also overcast the images, thus not providing the proper contrast to determine the boundary of leave or canopy from which further shape features need to be extracted (Tian et al., 2000). The insect/pest attack or leave disease may also cause the geometrical irregularities in structure thereby influencing the classifying range of different shape based parameters. These limitations served as bottlenecks for real-time application ultimately leading towards the exploration of new techniques to describe the plant texture by undermining its botanical information (Meyer, 2011). These texture based weed-crop segmentation approaches have been widely adapted by the researchers as a tool for variable rate applications (Shearer and Holmes, 1990; Meyer et al., 1998; Burks et al., 2000; Tang et al., 1999, 2003; Burks et al., 2005; Ghazali et al., 2007; Siddiqi et al., 2009; Bossu et al., 2009; Kiani and Kamgar, 2011; Ahmad et al., 2011; Chang et al., 2012; Ahmed et al., 2014; Kumar and Prema, 2016).

The texture of underlying terrain was first quantified by statistically estimating the probability of spatial distribution of an image pixel and its neighboring tonal variations at different orientations (Haralick et al., 1973). An array of fourteen second order statistical features extracted from "spatial gray-tone dependence matrices" was used for aerial photographs and satellite imagery with accuracies of 82% and 83%, respectively (Haralick et al., 1973). Shearer and Holmes (1990) implemented the same idea on HSI color images for classifying the different cultivars of nursery stock with an accuracy of 91%. Three different matrices were developed for each color plane (H, S and I) and 33 textural features as suggested by Haralick et al. (1973) were extracted by repeating the procedure for each color plane. Meyer et al. (1998) also used the same matrices for gray scale images of plants and soil with a reduction of features to four. The results showed that canonical discriminant analysis best identified the differences between different classes.

Burks et al. (2000) developed a weed discrimination system depending on color co-occurrence matrices (CCMs) followed by statistical discriminant analysis and stated an overall accuracy of 93%. Burks et al. (2005) compared the performance of four different classification techniques (statistical discriminant analysis, counter-propagation, backpropagation and radial bias neural network) by feeding the textural features extracted from CCMs as an input. The results of this evaluation revealed that the backpropagation algorithm outperformed the others by correctly classifying 97% of the images included in the study. In addition to the HSI color space Chang et al. (2012) used the concept of CCM to discriminated weeds from wild blueberry. The highest classification accuracy of reduced features (94.9%) was achieved by HSI color space. The addition of the luminance did not show the promising results.

Different transform based techniques characterize the multi-scale textural feature of the weed/crop by treating their images as two dimensional modulating frequency with different spatial dimension and orientation. Tang et al. (2003) introduced a filter window function to Fourier transform at a fixed orientation. The translation of the filter window across the entire image resulted into the different textural features. The filtering operation was performed on the signals of green channel with an intensity modulating between zero to nine with different window size. Ishak et al. (2009) improved the efficiency of the Gabor wavelet by adapting the gradient field distribution and curve fitting approach. The intensity, dimension and the orientation of Gabor wavelet were fixed while the gradients distribution of gradient field distribution algorithm was rotated according to the leave direction. The results showed an accuracy of 93.7%. The analysis of other studies on the transform based textural analysis indicated that the Daubechies wavelet still has a greater potential to explore (Okamoto et al., 2007; Bossu et al., 2009). A relatively new approach used a combination of textures extracted from the wrapping based culvert transform at an intensity level of 2 and 5 to capture both coarse and fine textures of the image along with a set of tamura features (Kumar and Prema, 2016). Though they achieved very high accuracy of 99% with random vector machines, the computational complexity of this algorithm, however, still needs to be evaluated.

The statistical and transform based textural approaches have been widely used and studied for weed-crop segmentation. However, there is relatively less research done towards the model based application and using structural descriptors (Tuceryan and Jain, 1998) to define the agricultural textures. Moreover, no research has been thoroughly done on the gray level run length matrices and normalized gray tone difference matrices to statistically undermine the texture of weed-crop images. While these approaches have shown their good results in other agricultural applications (Tahir et al., 2007).

Table 4. Comparison of different statistical and transform based texture analysis techniques for weed-crop segmentation.

Algorithm	Selected Features	Crop	Classification Technique	Best Classifying Features	Maximum Classification Accuracy	Reference
Color co-occurrence matrices	Angular second moment, mean intensity, variance correlation, product moment, inverse difference moment, entropy, sum entropy, difference entropy, information measures of correlation I and II	Seven different weeds found in nursery stock	Quadratic discriminant analysis	Six different feature models were selected using stepwise discriminant analysis	90.9%	Shearer and Holmes, 1990
Gray level co-occurrence matrices	Angular second moment, inertia, entropy and local homogeneity	A database of broad and narrow leave weed species	Linear and canonical discriminant analysis	Local homogeneity, angular second moment, Inertia	96.7% for soil	Meyer et al., 1998
Color co-occurrence matrices	Angular second moment, mean intensity, variance correlation, product moment, inverse difference moment, entropy, sum entropy, difference entropy, information measures of correlation I and II	Five weed species common in Kentucky row crops	Linear discriminant analysis	Six different feature models were selected using stepwise discriminant analysis	91.7%	Burks et al., 2000

Gabor wavelet	Features extracted by convolving a filter-bank with intensity levels ranging from 0–9 and dimension of 9 × 9, 13 × 13 and 17 × 17 along the green channel output at an orientation of 90°	A database of broad and narrow leave weed species	Backpropagation artificial neural network	Filters with intensity level of 4–7 and spatial dimension of 17 × 17 at an orientation of 90°	99.67% for grasses and 99.54% for broad leave weeds on few images	Tang et al., 2003
Color co-occurrence matrices	Angular second moment, mean intensity, variance correlation, product moment, inverse difference moment, entropy, sum entropy, difference entropy, information measures of correlation I and II	Five weed species common in Kentucky row crops	Linear discriminant analysis, backpropagation neural network, counter-propagation neural network, radial basis function	Six different feature models were selected using stepwise discriminant analysis	97% with backpropagation neural network	Burks et al., 2005
Daubechies wavelet at level of 4	240 approximation coefficients extracted from DB4	Sugar beet	Euclidian distance and Linear discriminant analysis	50 highest approximation coefficients using stepwise discriminant analysis	81.3% with linear discriminant analysis	Okamoto et al., 2007
Comparison between multi-level wavelet decomposition and Gabor filtering	33 different decomposition level of Daubechies, Symlet, Coiflet, Biorthogonal, Reverse biorthogonal and Discrete from of Mayer	Wheat	Empirically defined threshold levels	Daubechies-25 and Mayer	89.7% for Daubechies-25	Bossu et al., 2009

Table 4 cont. ....

...Table 4 cont.

Algorithm	Selected Features	Crop	Classification Technique	Best Classifying Features	Maximum Classification Accuracy	Reference
Gabor wavelet with gradient field distribution	Features extracted by convolving a filter at a frequency of 6.5 Hz at an orientation of 45° with a mask of 150 × 150 spatial dimension along the green channel followed by gradient orientation computation	Palm oil plantation	Single layer perceptron	Combined gradient vectors computed from Gabor wavelet in the first quadrant only	93.75%	Ishak et al., 2009
Multi-level wavelet decomposition	100 highest and average coefficients from each level of Daubechies, Symlets, Biorthogonal and reverse biorthogonal wavelets	Broad and narrow leaves weed database	Euclidian distances based classifier	100 highest and average coefficients from Symlet at a level of 4	98.31%	Siddiqi et al., 2009
Wavelet transformed images followed by Co-occurrence matrices	Energy, entropy, contrast, homogeneity and inertia from DB4 wavelet	Corn	Multi-layer perceptron neural network	No feature selection procedure was carried out	99.9% for corn and 96% for weeds	Kiani and Kamgar, 2011
Haar Wavelet transform	200 horizontal, vertical, diagonal and approximation coefficients extracted during decomposition stage	Broad and narrow leaves weed database	K-nearest neighbor	No feature selection procedure was carried out	94% for broad leave and 92% for narrow leave weeds	Ahmad et al., 2011

Color co-occurrence matrices	Angular second moment, contrast, variance correlation, product moment, inverse difference moment, entropy, sum entropy, difference entropy, information measures of correlation I and II	Wild blueberry	Multiple discriminant analysis	Three algorithms with stepwise feature selection	100% for bare spots, 98.2% for wild blueberry and 93.5% for weeds	Chang et al., 2012
Local binary pattern, Local ternary pattern and local directional pattern	Different patterns extracted by adjusting the range of parameters	A database of broad leave grass found majorly in the fields of Bangladesh	Template matching and support vector machines	No feature selection procedure was carried out	98.5% for local directional pattern with support vector machines	Ahmed et al., 2014
Wrapping based culvert transform with Tamura texture feature	Mean value and standard deviation of culvert at level 2 and 5 along with coarseness, contrast, directionality, line-likeness, regularity, roughness, energy, entropy and auto correlation	Eggplant	Relevance vector machines	No feature selection procedure was adapted	99%	Kumar and Prema, 2016

3. APPLICATION OF SMART CAMERAS FOR SORTING OF FRUITS, VEGETABLES AND GRAINS

The relatively more reliable and subjective approach of inspecting the agricultural products enables the machine vision based grading systems to outperform their human counterparts (Cubero et al., 2011). Moreover, the high throughput capacity (Aleixos et al., 2002) and less involvement of human factors (Cubero et al., 2011) forced the commercial processing plants to opt for these systems (Du and Sun, 2006). The rapidly fostering awareness about the food quality over the last two decades resulted in an evolution of the machine vision based grading for different agricultural products (Aleixos et al., 2002; Blasco et al., 2003; Leemans and Destain, 2004; Tahir et al., 2007; Jarimopas and Jaisin, 2008; Blasco, 2009a,b; Liming and Yanchao, 2010; Guevara-Hernandez and Gomez-Gil, 2011). Also the climatological effects on the physiological shape and structure of agricultural products invoked the interest of engineers, researchers and scientists across the world to develop custom grading systems according to the local needs and demands.

The external quality of both fresh and processed agricultural products can be defined by their shape, weight, color and presence of any blemishes or diseases (Cubero et al., 2011). Machine vision systems can correlate these external attributes of agricultural produce to its color, shape, size, volume and/or textural features (Fig. 3) and can aid in improving the efficiency of the grading process (Du and Sun, 2006). The different grading systems for fruits, vegetables and grains that emerged during the last two decades on the basis of these attributes are summarized in the next sections. The following sub-sections are categorical review of different grading systems and their targeted agricultural commodity.

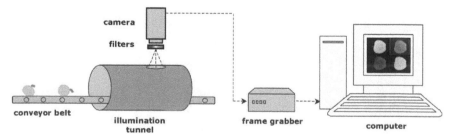

Fig. 3. Illustration of the image acquisition system used (Reprinted from Computers and Electronics in Agriculture. 75, Unay et al., Automatic grading of bi-colored apples by multispectral machine vision. Computers and Electronics in Agriculture, 204–212. Copyright (2011), with permission from Elsevier).

3.1 Color Information for Grading

Color information can help to identify the current ripening stage of agricultural produce which in turn can be used as a quality predicator for fruit grading systems. Blasco et al. (2009b) used simple pixel-oriented algorithm by removing the background from pomegranate arils using an adjustable threshold switch for a Red channel. In addition to the color information, size and centroid information was used to separate out the large/small unwanted material. A similar approach was used to grade apples on the basis of their color features and organizing feature parameter algorithm (OFP) (Xiaobo et al., 2007). Mean and variance features from RGB color space along with the definition of hue between 0° to 80° were used to categorize apples into four classes. The results showed that OFP algorithm outperformed the BPNN with slightly lower performance than SVM.

A comparison between five different color spaces (RGB, HSI, LUV, Lab and XYZ) was performed to identify the citrus peel defects (Blasco et al., 2007b). The classification criterion on the basis of LDA indicated that all color spaces achieved good (> 80%) accuracy except XYZ color space. A strawberry grading system used the "a" coordinate of Lab color space with empirically selected threshold levels to categories the fruit into black-red, bright-red and light-red (Liming and Yanchao, 2010). An apple grading system used average color values (R, G and B), variances (Vr, Vg and Vb) and chromatic color values (r, g and b) with two neural networks to identify the apples on the basis of the percentage of red color (Nakano, 1997). The grade judgment ratios of two out of five classes were found to be very low (65% and 32%), while the highest and mean judgment ratios were 95% and 70%, respectively. López-García et al. (2010) used RGB color data extracted from multi-resolution square window to define the reference Eigen-space. The reference space was used to build the PCA based model of pixel locations belonging to non-defected areas. The system was able to achieve a performance accuracy of 91.5% for different surface defects and 100% for stem end detection.

The color based grading algorithms were also used for different vegetables including bell peppers, olives, mushrooms, tomatoes and potatoes. Shearer and Payne (1990) mapped the hue component for different primary and secondary colors to identify the different colored and defected bell peppers. Feature selection procedure and quadratic discriminant functions were developed to define the sets of optimum features for individual type of defect. A set of three statistical features, along with range and edges were extracted from the raw RGB, normalized RGB and intensity color space to identify the potato blemishes (Barnes et al., 2010). The adaptive boost (AdaBoost) classifier was able to correctly identify the 89.6% and 89.5% blemishes on the white and red potatoes, respectively. Red band was used to calculate the mean reflectance spectra from 3 × 3 ROI, region of interest, from the center part of the mushroom surface to identify the defects caused by freezing and thawing (Gowen et al., 2009). PCA and LDA were able to correctly identify 76.2% of defected mushrooms. A color and color homogeneity descriptors were used to categories the tomato (Laykin et al., 2002). Mean and standard deviation of RGB channels, hue estimated from 40 × 40 pixels and average color on the basis of Quad Tree method were used to describe the color. The color homogeneity was estimated by dividing the fruit using virtual elliptical rings. The statistics including mean, median, mode and standard deviation of these individual rings were compared. The results showed that their system achieved 92% correct color homogeneity classification and 90% correct color detection. A standalone color based algorithm was used to differentiate between Canada western red spring (CWRS) wheat, Canada western amber durum (CWAD) wheat, barley, oats and rye with the help of LDA and K-nearest neighbor (K-NN) classifiers(Majumdar and Jayas, 2000a). The accuracies of 94.1, 92.3, 95.2 and 92.5%, were reported for the CWRS wheat, CWAD wheat, barley, oats and rye, respectively..

3.2 Shape and Size Estimation

In addition to the color information reflected by particular agricultural produce, size is also a major contributor to decide about its commercial fate. The size of the fruit in combination with its color can help the machine vision based sorting device to grade it according to the predetermined categories (Cubero et al., 2011). A size based apple grading system codified the extracted boundary of fruit using chain code to estimate the area and fruit size (Blasco et al., 2003). Additionally, the major damage length and total damage are also determined to avoid the influence of damages on final fruit grade. Leemans and Destain (2004) used the combination of size, color, texture and position attributes to correctly classify the apple fruit into two commercial grades. The results of their study showed a correct global classification rate of 73% from the combination of all these features. Unay et al. (2011) used perimeter as an estimator of size along with the circularity to define shape and defect ratio as a descriptor of defected surfaces along with other color and textural features to grade the apples in two and multiple classes. In addition to the linear models, several multi-factor non-linear approaches were also tested to achieve precise measurement of the defects and minimum confusion with calyx/ stem parts. The results of study showed that two-class grading approach achieved more appropriate results compared to their multi-grade counterpart.

Aleixos et al. (2002) used convolution mask, contour extraction and singularization as preprocess steps before extracting the geometrical features to define the fruit size. The shape of the fruit is estimated by using the circularity and relating the maximum diameter of the fruit. In addition to these parameters, external defects were also used to further guide the sorting device. The accuracy of the complete system was up to 94% for mandarins and 93% for lemons. The shape based descriptors are also used to categories the individual satsuma (mandarin) segments into whole and broken one for the real-time in-field applications at the sorting facility (Blasco et al., 2009a). The system was also capable of identifying the pieces of skin and other raw materials present in the segment batches by using simple Bayesian discriminant analysis approach. The shape of the segments was defined by using circularity, compactness, symmetry, elongation and Fourier descriptors, while the size was described using area and length of individual segment. Another real-time complete sorting solution was designed for the dates on the basis of their shape estimation and skin delamination criterion (Lee et al., 2008). The connected component analysis was performed to identify the presence of the date on the conveyor belt followed by the estimation of the fruit size either using length or area. The system showed an accuracy of 95% for Jumbo date.

The shape of strawberry was described by using the sharing line method into long-taper, square, taper and rotundity (Liming and Yanchao, 2010). Each strawberry was divided evenly using a set of seven horizontal and vertical lines with a condition of passing first line pair from the gravity center of the fruit. The difference between consecutive lengths was used as a descriptor of the fruit shape. The daimeter of the same fruit was estimated by using maximum horizontal line length. The system was able to correctly identify 90% of the shape in worst scenarios. Jarimopas and Jaisin (2008) used the curvature of the sweet tamarind pods to define the curved, slightly curved and straight shape based categories. The curvature was estimated by using a circle of 55-pixel radius in counter clockwise direction. This curvature was further used to draw a graph to locate the pulses indicating the presence of stem and tail of tamarind pod. The length of the pod was also estimated to with a maximum accuracy of 94.3%.

3.3 Textural Traits

Color and shape attributes can play major role in grading the fruit and vegetable, however, they alone may not be able to help in grading of the different grains because of the very similar color reflectance and shape

Table 5. Summary of different machine vision inspection systems for fruits.

Agricultural Product	Sorting Criterion	Characterizing Features	Classification Technique	Classification Accuracy	Reference
Apple	Surface color based grading	Average color coordinates, variances and normalized color coordinates from RGB colors	Artificial neural network	95% for superior (AA) and 32% for excellent (A) grade quality	Nakano, 1997
Apple, Oranges and Peaches	Size, color, stem location and external blemishes	Primary color averages, secondary color averages, fruit area, fruit size, fruit centroid, length of major damage, Total damage area, stem and calyx centroid	Bayesian discriminant analysis	99%, 95% and 80% peel detection and 87%, 100% and 89% defects detection for apples, oranges and peaches, respectively.	Blasco et al., 2003
Apple	Skin defects detection	Mean values and standard deviations of RGB, Mean gradient and standard deviation gradient of red channel, Euclidean distance between the fruit background and defect mean color, fourth root of area, square root of perimeter, major inertia moment, ratio of inertia moments, centroid	K-mean clustering	73%	Leemans and Destain, 2004

Crop	Application	Features	Classifiers	Results	References
Apple	Color based decision system	Average color gradients, variances and color coordinates from RGB color space and eight features from Hue color space	Organizing feature parameter, back propagation artificial neural network and support vector machines	85%, 80.5%, 72.5% and 88.2% for extra, class-I, class-II and rejected classes respectively on the basis of test data	Xiaobo et al., 2007
Apple	Blemishes detection	Flooding algorithm for coarse detection of blemishes followed by a close loop snake algorithm for defects localization	A threshold of 2 or more defects per apple as a rejection criterion	89%	Xiaobo et al., 2010
Apple	Surface defects including bruising, rot, small russet and scar tissues	Mean, standard deviation, median, minimum, maximum, perimeter, circularity, defect ratio, angular second moment, contrast, variance, inverse difference moment	Linear discriminant analysis, K-nearest neighbor, fuzzy K-nearest neighbor, support vector machines, syntactical classifiers	93.5% for two class grading system	Unay et al., 2011
Apple and Guava	Bruise detection for apple and skin damage for guava	3D volume reconstruction followed by the normalization, mean and standard deviation extracted from binary, RGB and HSI color models along with color co-occurrence matrices and run-length textural features extracted from RGB and HSI	K-nearest neighbor, artificial neural network, support vector machines, genetic programming	95.7% for apple with ANN and 95.1% with genetic programming for guava	Yimyam and Clark, 2016

Table 5 cont. ....

...Table 5 cont.

Agricultural Product	Sorting Criterion	Characterizing Features	Classification Technique	Classification Accuracy	Reference
Oranges, Mandarins and lemons	Defects detection and size estimation	Centroid, maximum and minimum diameters, perimeter and circularity along with the Bayesian look up table developed from RGBI	Bayesian discriminant model	93% for lemon and 94% for mandarins	Aleixos et al., 2002
Citrus	Surface defects detection including disease and mold	Color values from RGB, HSI, Lab, LUV, XYZ, NIR, FL and UV images	Linear discriminant analysis	87.2%, 83.3%, 83.7%, 82.1%, 71.1%, 63.4%, 79.4% and 92.9% for HSI, Lab, RGB, LUV, XYZ, FL, UV and NIR respectively	Blasco et al., 2007a
Citrus	Peel defects detection	Region growing with a window size of 3 × 3 and smoothing using Euclidean distance	Threshold defined on the basis of mean and standard deviation of the pixel colors in the selected windows	94.2% total	Blasco et al., 2007b
Satsuma (Mandarin) segments	Segments physical condition	Area, axis of inertia, roundness, elongation, compactness, symmetry, 10 harmonics of FFT of shape signature	Non-linear Bayesian discriminant analysis	93.2% of complete fruit segments	Blasco et al., 2009a
Citrus (Oranges and Mandarins)	Surface defects including damage and disease infestation	Multi-resolution unfolded matrix of raw RGB data used to define the reference Eigen-space	Principal component analysis	91.5% for individual defects and 94.2% for classification	López-García et al., 2010

Date	Skin delamination	Connected component analysis for area or length estimation along with erosion and Sobel operator application for edge detection	User selected threshold values for detecting skin delamination	95% for Jumbo date	Lee et al., 2008
Date	Grading on the basis of physiological attributes and defects detection	Flabbiness estimated from the RGB values, area as an estimator of size, Fourier coefficients for relating the edges extracted from Sobel Operator	Back propagation artificial neural network	85%	Al Ohali, 2011
Pomegranate arils	Detection of different foreign materials including rotten arils and membrane pieces	Segmentation of arils in Red channel followed by the size and centroid estimation to remove extra small and large objects	Manually selected thresholds Linear discriminant analysis	Over 90% with both methods	Blasco et al., 2009b
Strawberry	Grading on the basis of shape, color and texture	Lab color space for color, edge detection and sharing line algorithm to define shape horizontally and vertically, maximum horizontal line to relate the image diameter and actual diameter of fruit	Manually defined thresholds with K-means clustering	88.8% for color, 90% for shape and above 90% for size except one class	Liming and Yanchao, 2010
Sweet Tamarind	Shape, Size and defects detection	Shape index developed by recording the radial coordinates, length of tamarind pod and color based crack detection	Threshold levels defined on the basis of analysis of variance	94.3% and 89.8% for two different cultivars	Jarimopas and Jaisin, 2008

Table 6. Summary of different machine vision inspection systems for vegetables and grains.

Agricultural Product	Sorting Criterion	Characterizing Features	Classification Technique	Classification Accuracy	Reference
Bell pepper	Grading by color and defects	Hue mapping of Orange, yellow, cyan, green, blue, violet, magenta and red color	Quadratic discriminant analysis	96% by color and 63% by defects detection	Shearer and Payne, 1990
Mushrooms	Defects caused by freezing	Mean red spectral reflectance for 10 regions of interest and L channel of Hunter Lab color space	Principal component analysis, linear discriminant analysis	76.2% for freezing damage on test data	Gowen et al., 2009
Olives	Differentiation between tree and ground olives along with wrinkles detection	Gradient images for wrinkle identification and difference of color images were treated with filter algorithm and PCA	Linear discriminant analysis and multi-layer perception neural network	100% with multi-layer perceptron neural network	Puerto et al., 2015
Olives	Surface defects including scratches, stain, pitting, split and hail damage	Pixel number of lighter skin, darker skin, light defect, dark defect and unusual dark color	Bayesian discriminant analysis, back propagation artificial neural network and partial least squares multi-discriminant analysis	86% for Bayesian, 100% for BP-ANN and 90% for PLS	Diaz et al., 2004
Potato	Blemishes detection	Mean, variance, skew, edges, color and range features extracted from raw RGB values, normalized RGB color values and Intensity	Minimalist adaptive boosting algorithm	89.6% and 89.5% for white and red potato, respectively	Barnes et al., 2010

Tomatoes	Defects, shape, color, color homogeneity and stem detection	Color estimation using standard deviation, side block and quad tree method, Sobel edge in hue image with elliptical rings for calculating mean, median and mode to define color homogeneity, comparing counted stem pixels for stem identification, FFT transform radii sequence of tomato outline to define roundness and adaptive threshold in red image for defects detection	Manually defined several thresholds to estimate the individual parameter	90% bruise, 90% color homogeneity, 92% color and 100% stem detection	Laykin et al., 2002
CWRS wheat, CWAD wheat, barley, oats, and rye	Color based grading	Mean, variance, and range of red, green, blue, hue, saturation, and intensity.	Linear discriminant analysis, K-nearest neighbor	94.1, 92.3, 93.1, 95.2, and 92.5%, for CWRS wheat, CWAD wheat, barley, oats, and rye respectively	Majumdar and Jayas, 2000a
CWRS wheat, CWAD wheat, barley, oats, and rye	Shape and size based grading	Area, Perimeter, length, width, major axis length, minor axis length, thinness ratio, aspect ratio, rectangular aspect ratio, area ratio, maximum radius, minimum radius, standard deviation of all radii, Haralick ratio, Fourier descriptors and spatial moments	Linear discriminant analysis, K-nearest neighbor	98.9, 93.7, 96.8, 99.9, and 81.6%, respectively for CWRS wheat, CWAD wheat, barley, oats, and rye	Majumdar and Jayas, 2000b
CWRS wheat, CWAD wheat, barley, oats, and rye	Texture based grading	Mean, variance, uniformity, entropy, maximum probability, correlation, homogeneity, inertia, cluster shade, cluster prominence, short run, long run, gray level non-uniformity, run length non-uniformity, run percent. GLRM entropy	Linear discriminant analysis, K-nearest neighbor	85.2, 98.2, 100.0, 100.0, and 76.3%, for CWRS wheat, CWAD wheat, barley, oats, and rye respectively	Majumdar and Jayas, 2000c

Table 6 cont. ...

...Table 6 cont.

Agricultural Product	Sorting Criterion	Characterizing Features	Classification Technique	Classification Accuracy	Reference
Wheat and Barley	Classification according to the moisture content	Short run emphasis, long run emphasis, gray level non-uniformity, run length non-uniformity, low gray level run emphasis, run percentage and high gray GLRM for angles 0°, 90°, 45° and 135°. Mean, variance and ranges of RGB and HSI, area, perimeter, major axis length, minor axis length, maximum radius, minimum radius, mean radius, four invariant shape moments and 20 harmonics of Fourier descriptors (FD) and features extracted from GLCM	Linear discriminant analysis, K-nearest neighbor, BP-ANN	Up to 98% for bulk grain samples and up to 68% for the individual grain samples	Tahir et al., 2007
Wheat and Barley grains	Identification of two grain types on the basis of their external characteristics	Area, perimeter, thinness ratio, length, width, equivalent rectangle long side, equivalent rectangle short side, ratio of equivalent rectangle sides, the seven Hu moments, rectangular aspect ratio, maximum radius, minimum radius, radius ratio, radius standard deviation, Haralick radius, Mean and standard deviation of RGB, mean and variance of XY, uniformity, maximum probability, correlation homogeneity, cluster shade and cluster prominence of the GLCM for angles 0°, 90°, 45° and 135°, short run emphasis, long run emphasis, gray level non-uniformity, run length non-uniformity, low gray level run emphasis, run percentage and high gray 0°, 90°, 45° and 135°	Discriminant analysis and K-nearest neighbor	99% by combining the properly selected shape, color and texture features	Guevara-Hernandez and Gomez-Gil, 2011

properties. One reason could be the small size of grains which could not be precisely estimated resulting into poor categorization. Furthermore, the variability in the moisture content can cause the grains to shrink or swell resulting in the change in size (Tahir et al., 2007). Therefore, different statistical descriptors in combination with shape and color information were used to quantify the texture of grain items, thereby helping in the sorting operation. Majumdar and Jayas (2000c) used textural features extracted from the gray level co-occurrence matrices (GLCM) and grey level run length matrix (GLRM) to identify between CWRS wheat, CWAD wheat, barley, oats and rye grains. The parametric and non-parametric methods were used to develop the identifying models and best results were achieved using K-nearest neighbor with a level of k = 5.

Tahir et al. (2007) used the combination of color, shape and textural features to quantify the effect of moisture content on the grain kernel morphology and appearance. The images of CWRS, CWAD and barley were taken in individual and bulk fashion from the conditioned grains with moisture content varying from 12% to 20%. It was observed that the highest contribution towards the identification was from color followed by the textural features. Guevara-Hernandez and Gomez-Gil (2011) used a similar technique to classify wheat and barley kernels with discriminant analysis and K-NN. The GLCM and GLRM were developed in four different orientations and the similar features were extracted from different orientations. The authors concluded that the combination of shape, color and texture can provide the better accuracy as compared to any of these individually. The classification accuracy can be as high as 99% by carefully selecting a set from these pooled features (Guevara-Hernandez and Gomez-Gil, 2011).

4. HARDWARE BASED IMAGE PROCESSING TOOLS—A WAY FORWARD TO MINIMIZE THE COMPUTATIONAL EXPENSES

The application of sophisticated modern image processing algorithm demands a very high-end computationally efficient central processing unit for their real-time applications. Currently and most commonly used computational platforms in agricultural sector are based on personalized computers, because of the relatively easier image processing programming needed for them. These devices, however, were not able to match the processing speed needed for the real-time applications (Lee et al., 1999; Chang et al., 2012) and therefore limit the travel speed of the platform. Chang et al. (2012) concluded that using a personalized computer, a compromise between the accuracy of CCM algorithm for weed-crop segmentation and processing time is needed. The authors were able to achieve a high accuracy of 94.9%, but a travel speed of only 3.1 km hr^{-1}

inhibits its real-time application. Aleixos et al. (2002) reported that use of traditional sorting systems can only handle the tasks requiring less computational time, thereby reducing the overall accuracy of the systems to achieve the required commercial grade sorting speed. In contrast to the traditional PCs, the hardware based embedded solutions including DSP, FPGA, ARM and GPU proved their capabilities for real-time machine vision based agricultural applications (Murphy et al., 2007; Pearson, 2009; Pearson, 2010; Teixidó et al., 2012; Pearson et al., 2013; Singh, 2014; Mohan et al., 2016). The minimized computational time using hardware based embedded solutions can help to achieve higher accuracy and higher speed simultaneously with existing algorithms for real-time applications. The reduction in computational expenses can also help to include more features thereby allowing to craft more complex multi-feature algorithms for increased accuracy without having any limitations of delay caused by slow processing speed.

A relatively less number of studies have been reported for FPGA based embedded systems in agricultural sector because of the much more challenging design and programming complexities. Murphy et al. (2007) implemented a fairly simple census transform algorithm on the grayscale images of two cameras for the FPGA based stereo-vision system. A color features based high speed sorting system was developed by raising a FPGA circuitry board directly on the image sensor board (Fig. 4). The system (Fig. 5) was responsible for separating the white wheat from red one and inspecting the popcorn for blue-eye damage. The throughput capacity of proposed system was 8 kg per hour of wheat and 40 kg per hour of popcorn. The detection accuracy for wheat was acceptable (88%–90%), while for the popcorn it was low (74%). A relatively more complex image processing algorithm on the basis of color features and LDA was implemented on a prototype grain sorter (Pearson, 2010). An advanced FPGA board containing more logical elements and memory compared to Pearson (2009) was used to perform more rigorous tasks. This system achieved a higher throughput capacity (25 kg per hour) with information being processed more accurately. Similar results were reported by the other studies (Pearson et al., 2012; Pearson et al., 2013), however, none of them have reported the application of emerging algorithms on FPGAs.

A multispectral machine vision system was developed using a pair of DSPs for inspecting and sorting the oranges (Aleixos et al., 2002). The system implemented a master/salve configuration of DSP devices for connecting two cameras. The master DSP extracted the geometrical features from a monochrome camera fixed with a NIR filter along with the salve DSP responsible for detecting the skin damages using RGBI

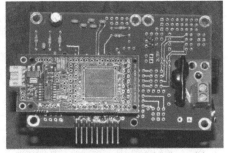

Image sensor side with lens

FPGA board side, the FPGA board is the smaller blue board raised above the image sensor board.

Fig. 4. Photo of the image sensor and FPGA boards connected together. The image sensor and lens are on the opposite side from the FPGA board (Reprinted from Computers and Electronics in Agriculture. 69(1), Pearson, Hardware-based image processing for high-speed inspection of grains, 12–18. Copyright (2009), with permission from Elsevier).

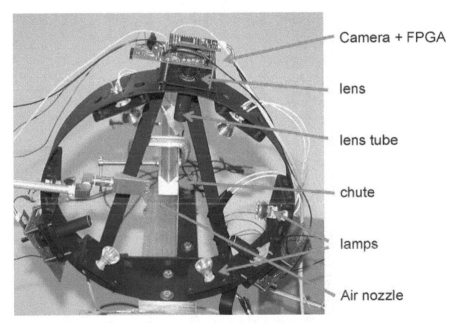

Camera + FPGA

lens

lens tube

chute

lamps

Air nozzle

Fig. 5. End view of the sorter sensing system showing all three cameras, six light bulbs, air nozzle, and chute (Reprinted from Computers and Electronics in Agriculture. 69(1), Pearson, Hardware-based image processing for high-speed inspection of grains, 12–18. Copyright (2009), with permission from Elsevier).

Table 7. A review of different hardware based embedded systems implemented for agricultural vision applications.

Processor Used	Camera Used	Application	Features	Complexity Level	Reference
Master and slave DSP units working on shape and color features separately	A color CCD camera and a monochromatic CCD camera with infrared cut filter	Orange sorting system using hyperspectral cameras	Combination of geometrical/shape and color features	High	Aleixos et al., 2002
FPGA (Xilinx Spartan-3 XC3S2000)	Omnivision CMOS color image sensors	Low cost stereo-vision for agricultural application	Census transform algorithm	Low	Murphy et al., 2007
Pluto-II FPGA board (Altera Cyclone EP1C3)	CMOS color image sensor (KAC-9628)	High-speed inspection of wheat and popcorn	Color features followed by a decision made on fixed global thresholds	Low	Pearson, 2009
Pluto-III FPGA board (Altera Cyclone EP2C5)	CMOS color image sensor (KAC-9628)	Sorting of wheat kernels	Color based surface texture features along with linear discriminant analysis	Intermediate	Pearson, 2010
FPGA (Altera Cyclone EP2C20Q240C8)	CMOS color image sensor (KAC-9628)	Detection of small spots on popcorn	Color features along with the empirically defined three threshold levels	Intermediate	Pearson et al., 2012
ARM Cortex-M4 (STM32F407VGT6)	Omnivision (OV7670) color image sensor	Red peach detection system	Three dimensional LUT in RGB color space	Intermediate	Teixidó et al., 2012

Pluto-III FPGA board (Altera Cyclone EP2C5)	CMOS color image sensor (KAC-9628)	Fungal damaged white sorghum detection	Color based features along with discriminant analysis	Intermediate	Pearson et al., 2013
FPGA (Xilinx Spartan-3E XC3S250E)	Ordinary color camera	Detection of greening of potatoes	Simple color features along with manually defined threshold	Low	Singh, 2014
LPC2148 (ARM-7 based micro-controller)	Ordinary webcam	Weed detection system at laboratory scale	Color features followed by a decision made on fixed global thresholds	Low	Mohan et al., 2016
NVIDIA TitanX	Not specified	Plant identification	Object detection based on Deep nets (Jia et al., 2014)	High	Lee et al., 2016
NVIDIA Tesla K40 for training, GeForce GTX 980M	JAI multi-spectral camera (RGB + NIR) and Microsoft Kinect2	Fruits (pepper, melon and apple) detection	Multimodal faster Region-based Convolutional neural network (R-CNN)	High	Sa et al., 2016
GeForce GTX TITAN X 12 Gb	Not specified	Leaf disease	Convolutional neural network	High	Sladojevic et al., 2016
NVIDIA Quadro K600	Not specified	Plant classification	Convolutional neural network	High	Yalcin and Razavi, 2016

information. The processed image information from the salve processor was transferred to the master which further shared the final results with control computer. This system was able to inspect the color, size and presence of the skin defects at a minimum rate of 5 fruits per second with the accuracy of 94%. Another embedded peach detection system used ARM Cortex™ based processor to create the 3D look up tables (LUTs) from both linear combinations and histograms of RGB color vectors. The system showed a least performance of 77% by correctly identifying the red peaches in orchards with occluded leaves.

The DSP, ARM and FPGA have the ability to tackle complex machine vision instructions because of parallel information handling approach compared to their sequential counterparts. However, the complication of programming these devices for texture and shape base analysis along with sophisticated decision making tools hampers their ability for the real-time application in agricultural sector. Therefore, programming gaps need to be filled in order to get the full advantage of their processing speed and accuracy.

Artificial neural network has been studied to process images but it needs large datasets which require tremendous human effort to collect, annotate and process to cover the full variability of the target (Guo et al., 2013). With a rapid progress of the GPU, deep neural nets [i.e., Deep Convolutional Neural Network (DCNN)] using the GPU have been recently introduced to overcome this constraint (Lee et al., 2016; Sa et al., 2016; Sladojevic et al., 2016; Yalcin and Razavi, 2016) in agricultural sector (plant identification, fruit detection, disease detection, etc.).

5. CONCLUSION

Since Thompson et al. (1991) showed the potential of weed detection for spot-application of herbicide, many different methods were used for real-time detection of weeds rather than manual surveying and remote sensing. A cost-effective smart camera which comprises of a machine vision system with application-specific algorithm using color, shape and texture has been used for weed-crop segmentation. Segmentation by shape has many limitations such as, same leaf shapes can occur on crops and weeds thus rendering the algorithm ineffective. During the early stage of weed emergence, shape based algorithms showed better results. Segmentation by texture is promising, however, there has been much less research done and also it can be more affected by outside factors especially from variability in outdoor lighting. Also it requires higher processing time due to complicated computation. Therefore, currently weed-crop segmentation by color with neural network based classifiers can be considered as one of the most suitable candidate for real-time applications.

A large number of studies have been done for agricultural product sorting using the cost-effective smart camera with color, shape and/ or texture based algorithms. When the cost-effective smart camera was used for sorting of agricultural products with several attributes, it showed positive results. Even though the wide variety of fruits and vegetables call for a wide variety of attributes, agricultural products sorting using the cost-effective smart camera are much better than human as they reduce human error and have a higher processing capacity. Even hyper-/ multi-spectral cameras were not discussed in this chapter. A cost-effective multispectral cameras using a double filter was used for plant detection (Dworak et al., 2013).

These days machine learning methods, especially supervised machine learning such as SVM, K-NN and artificial neural network, are used to increase the accuracy of smart camera. However, the number of training data set critically affects the accuracy of these methods because variability of the training data set needs to cover the full variability of the target (Guo et al., 2013) which requires tremendous human effort for classification. Recently, deep neural nets application including DCNN using the GPU is emerging as it extracts feature automatically which may help to reduce tedious manual efforts.

With extensive calculation burden for extracting shape and/or textural features from images, these algorithms need powerful rugged computer and/or a front look-a-head position which can ensure distance and enough processing time for calculation while increasing the complexity. The use of DSP, ARM, FPGA and/or GPU may reduce these processing time constraints.

REFERENCES

Ahmad, I., Siddiqi, M. H., Fatima, I., Lee, S., and Lee, Y. K. 2011. Weed classification based on Haar wavelet transform via k-nearest neighbor (k-NN) for real-time automatic sprayer control system. In Proceedings of the 5th International Conference on Ubiquitous Information Management and Communication (p. 17). ACM.

Ahmed, F., Kabir, M. H., Bhuyan, S., Bari, H., and Hossain, E. 2014. Automated weed classification with local pattern-based texture descriptors. Int. Arab J. Inf. Technol. 11(1): pp. 87–94.

Al Ohali, Y. 2011. Computer vision based date fruit grading system: Design and implementation. Journal of King Saud University-Computer and Information Sciences. 23(1): pp. 29–36.

Aleixos, N., Blasco, J., Navarron, F., and Molto, E. 2002. Multispectral inspection of citrus in real-time using machine vision and digital signal processors. Computers and Electronics in Agriculture. 33(2): pp. 121–137.

Andreasen, C., Rudemo, M., and Sevestre, S. 1997. Assessment of weed density at an early stage by use of image processing. Weed Research. 37(1): pp. 5–18.

Andújar, D., Escola, A., Dorado, J., and Fernández-Quintanilla, C. 2011. Weed discrimination using ultrasonic sensors. Weed Research. 51(6): pp. 543–547.

Andújar, D., Rueda-Ayala, V., Moreno, H., Rosell-Polo, J. R., Valero, C., Gerhards, R., and Griepentrog, H. 2013. Discriminating crop, weeds and soil surface with a terrestrial LIDAR sensor. Sensors. 13(11): pp. 14662–14675.

Bai, X., Cao, Z., Wang, Y., Yu, Z., Zhang, X., and Li, C. 2013. Crop segmentation from images by morphology modeling in the CIE L* a* b* color space. Computers and Electronics in Agriculture. 99: pp. 21–34.

Barnes, M., Duckett, T., Cielniak, G., Stroud, G., and Harper, G. 2010. Visual detection of blemishes in potatoes using minimalist boosted classifiers. Journal of Food Engineering 98(3): pp. 339–346.

Belbachir, A. N., and Göbel, P. M. 2009. Smart Cameras: A Historical Evolution. In Smart Cameras (pp. 3–17). Springer US.

Blasco, J., Aleixos, N., Cubero, S., Gómez-Sanchís, J., and Moltó, E. 2009a. Automatic sorting of satsuma (citrus unshiu) segments using computer vision and morphological features. Computers and Electronics in Agriculture. 66(1): pp. 1–8.

Blasco, J., Aleixos, N., Gómez, J., and Moltó, E. 2007a. Citrus sorting by identification of the most common defects using multispectral computer vision. Journal of Food Engineering. 83(3): pp. 384–393.

Blasco, J., Cubero, S., Gómez-Sanchís, J., Mira, P., and Moltó, E. 2009b. Development of a machine for the automatic sorting of pomegranate (punica granatum) arils based on computer vision. Journal of Food Engineering. 90(1): pp. 27–34.

Blasco, J., Aleixos, N., and Moltó, E. 2003. Machine vision system for automatic quality grading of fruit. Biosystems Engineering. 85(4): pp. 415–423.

Blasco, J., Aleixos, N., and Moltó, E. 2007b. Computer vision detection of peel defects in citrus by means of a region oriented segmentation algorithm. Journal of Food Engineering. 81(3): pp. 535–543.

Bossu, J., Gée, C., Jones, G., and Truchetet, F. 2009. Wavelet transform to discriminate between crop and weed in perspective agronomic images. Computers and Electronics in Agriculture. 65(1): pp. 133–143.

Burgos-Artizzu, X. P., Ribeiro, A., Guijarro, M., and Pajares, G. 2011. Real-time image processing for crop/weed discrimination in maize fields. Computers and Electronics in Agriculture. 75(2): pp. 337–346.

Burks, T., Shearer, S., Heath, J., and Donohue, K. 2005. Evaluation of neural-network classifiers for weed species discrimination. Biosystems Engineering. 91(3): pp. 293–304.

Burks, T., Shearer, S., and Payne, F. 2000. Classification of weed species using color texture features and discriminant analysis. Transactions of the ASAE. 43(2): pp. 441–448.

Camargo Neto, J. 2004. A combined statistical-soft computing approach for classification and mapping weed species in minimum-tillage systems.

Chaisattapagon, C., and Zhang, N. 1995. Effective criteria for weed identification in wheat fields using machine vision. Transactions of the American Society of Agricultural and Biological Engineers. 38(3): 965–974.

Chang, Y. K., Zaman, Q. U., Esau, T. J., and Schumann, A. W. 2014. Sensing system using digital photography technique for spot-application of herbicide in pruned wild blueberry fields. Applied Engineering in Agriculture. 30(2): pp. 143–152.

Chang, Y., Zaman, Q., Schumann, A., Esau, T., and Aylew, G. 2012. Development of color co-occurrence matrix based machine vision algorithms for wild blueberry fields. Applied Engineering in Agriculture. 28(3): pp. 315–323.

Cheng, H., Jiang, X., Sun, Y., and Wang, J. 2001. Color image segmentation: Advances and prospects. Pattern Recognition. 34(12): pp. 2259–2281.

Chi, Y., Chien, C., and Lin, T. 2003. Leaf shape modeling and analysis using geometric descriptors derived from Bezier curves. Transactions-American Society of Agricultural Engineers. 46(1): pp. 175–188.

Cubero, S., Aleixos, N., Moltó, E., Gómez-Sanchis, J., and Blasco, J. 2011. Advances in machine vision applications for automatic inspection and quality evaluation of fruits and vegetables. Food and Bioprocess Technology. 4(4): pp. 487–504.

Demirkaya, O., Asyali, M. H., and Sahoo, P. K. 2008. Image processing with MATLAB: Applications in medicine and biology. CRC Press.

Diaz, R., Gil, L., Serrano, C., Blasco, M., Moltó, E., and Blasco, J. 2004. Comparison of three algorithms in the classification of table olives by means of computer vision. Journal of Food Engineering. 61(1): pp. 101–107.

Du, C., and Sun, D. 2006. Learning techniques used in computer vision for food quality evaluation: A review. Journal of Food Engineering. 72(1): pp. 39–55.

Dworak, V., Selbeck, J., Dammer, K. H., Hoffmann, M., Zarezadeh, A. A., and Bobda, C. 2013. Strategy for the development of a smart NDVI camera system for outdoor plant detection and agricultural embedded systems. Sensors. 13(2): pp. 1523–1538.

El-Faki, M., Zhang, N., and Peterson, D. 2000. Weed detection using color machine vision. Transactions of the ASAE-American Society of Agricultural Engineers. 43(6): pp. 1969–1978.

Esau, T. J., Zaman, Q. U., Chang, Y. K., Schumann, A. W., Percival, D. C., and Farooque, A. A. 2014. Spot-application of fungicide for wild blueberry using an automated prototype variable rate sprayer. Precision Agriculture. 15(2): pp. 147–161.

Food and Agricultural Organization of United Nations (FAO). 2014. The State of Food and Agriculture 2015 IN BRIEF: Innovation in family farming. From http://www.fao.org/3/a-i4036e.pdf.

Food and Agricultural Organization of United Nations (FAO). 2015. How to Feed the World in 2050. From: http://www.fao.org/fileadmin/templates/wsfs/docs/expert_paper/How_to_Feed_the_World_in_2050.pdf.

Franz, E., Gebhardt, M., and Unklesbay, K. 1991. Shape description of completely visible and partially occluded leaves for identifying plants in digital images. Trans. ASAE. 34(2): pp. 673–681.

Franz, E., Gebhardt, M., and Unklesbay, K. 1995. Algorithms for extracting leaf boundary information. Transactions of the ASAE. 38(2): pp. 625–633.

Gebhardt, S., Schellberg, J., Lock, R., and Kühbauch, W. 2006. Identification of broad leaved dock (Rumex Obtusifolius L.) on grassland by means of digital image processing. Precision Agriculture. 7(3): pp. 165–178.

Gebhardt, S., and Kühbauch, W. 2007. A new algorithm for automatic Rumex Obtusifolius detection in digital images using colour and texture features and the influence of image resolution. Precision Agriculture. 8(1-2): pp. 1–13.

Gerhards, R., and Oebel, H. 2006. Practical experiences with a system for site-specific weed control in arable crops using real-time image analysis and GPS-controlled patch spraying. Weed Research. 46(3): pp. 185–193.

Ghazali, K. H., Mansor, M. F., Mustafa, M. M., and Hussain, A. 2007. Feature extraction technique using discrete wavelet transform for image classification. Paper presented at the Research and Development, 2007. SCOReD 2007. 5th Student Conference on, 1–4.

Golzarian, M. R., and Frick, R. A. 2011. Classification of images of wheat, ryegrass and brome grass species at early growth stages using principal component analysis. Plant Methods. 7(1): pp. 28.

Golzarian, M. R., Lee, M., and Desbiolles, J. 2012. Evaluation of color indices for improved segmentation of plant images. Transactions of the ASABE. 55(1): pp. 261–273.

Gonzalez, R., and Woods, R. 2008. Digital image processing: Pearson prentice hall. Upper Saddle River, NJ, USA.

Gowen, A. A., Taghizadeh, M., and O'Donnell, C. P. 2009. Identification of mushrooms subjected to freeze damage using hyperspectral imaging. Journal of Food Engineering. 93(1): pp. 7–12.

Grassini, P., Eskridge, K. M., and Cassman, K. G. 2013. Distinguishing between yield advances and yield plateaus in historical crop production trends. Nature Communications. 4.

Guerrero, J. M., Pajares, G., Montalvo, M., Romeo, J., and Guijarro, M. 2012. Support vector machines for crop/weeds identification in maize fields. Expert Systems with Applications. 39(12): pp. 11149–11155.

Guevara-Hernandez, F., and Gomez-Gil, J. 2011. A machine vision system for classification of wheat and barley grain kernels. Spanish Journal of Agricultural Research. 9(3): pp. 672–680.

Guijarro, M., Pajares, G., Riomoros, I., Herrera, P., Burgos-Artizzu, X., and Ribeiro, A. 2011. Automatic segmentation of relevant textures in agricultural images. Computers and Electronics in Agriculture. 75(1): pp. 75–83.

Guo, W., Rage, U.K., and Ninomiya, S. 2013. Illumination invariant segmentation of vegetation for time series wheat images based on decision tree model. Computers and Electronics in Agriculture. 96: pp. 58–66.

Guyer, D. E., Miles, G., Schreiber, M., Mitchell, O., and Vanderbilt, V. 1986. Machine vision and image processing for plant identification. Transactions of the ASAE. 29(6): pp. 1500–1507.

Guyer, D., Miles, G., Gaultney, L., and Schreiber, M. 1993. Application of machine vision to shape analysis in leaf and plant identification. Transactions of the ASAE. 36(1): pp. 0163–0171.

Haggar, R., Stent, C., and Isaac, S. 1983. A prototype hand-held patch sprayer for killing weeds, activated by spectral differences in crop/weed canopies. Journal of Agricultural Engineering Research. 28(4): pp. 349–358.

Hague, T., Tillett, N., and Wheeler, H. 2006. Automated crop and weed monitoring in widely spaced cereals. Precision Agriculture. 7(1): pp. 21–32.

Haralick, R. M., and Shanmugam, K. 1973. Textural features for image classification. IEEE Transactions on Systems, Man, and Cybernetics. 3(6): pp. 610–621.

Hamuda, E., Glavin, M., and Jones, E. 2016. A survey of image processing techniques for plant extraction and segmentation in the field. Computers and Electronics in Agriculture. 125: pp. 184–199.

Hemming, J., and Rath, T. 2001. PA—Precision agriculture: Computer-vision-based weed identification under field conditions using controlled lighting. Journal of Agricultural Engineering Research. 78(3): pp. 233–243.

Herrera, P. J., Dorado, J., and Ribeiro, Á. 2014. A novel approach for weed type classification based on shape descriptors and a fuzzy decision-making method. Sensors. 14(8): pp. 15304–15324.

Hu, M. 1962. Visual pattern recognition by moment invariants. IRE Transactions on Information Theory. 8(2): pp. 179–187.

Hunt, E. R., Hively, W. D., McCarty, G. W., and Daughtry, C. S. T. 2011. NIR-green-blue high-resolution digital images for assessment of winter cover crop biomass. GIScience and Remote Sensing. 48(1): pp. 86–98.

Ishak, A. J., Hussain, A., and Mustafa, M. M. 2009. Weed image classification using gabor wavelet and gradient field distribution. Computers and Electronics in Agriculture. 66(1): pp. 53–61.

International Labor Organization (ILO). 2014. Global Employment Trends 2014. From http://www.ilo.org/wcmsp5/groups/public/dgreports/dcomm/publ/documents/publication/wcms_233953.pdf.

Jarimopas, B., and Jaisin, N. 2008. An experimental machine vision system for sorting sweet tamarind. Journal of Food Engineering. 89(3): pp. 291–297.

Ji, R., Fu, Z., and Qi, L. 2007. Real-time plant image segmentation algorithm under natural outdoor light conditions. New Zealand Journal of Agricultural Research. 50(5): pp. 847–854.

Kataoka, T., Kaneko, T., Okamoto, H., and Hata, S. 2003. Crop growth estimation system using machine vision. Paper presented at the Advanced Intelligent Mechatronics, 2003. AIM 2003. Proceedings. 2003 IEEE/ASME International Conference on, 2 b1079–b1083 vol. 2.

Kazmi, W., Garcia-Ruiz, F. J., Nielsen, J., Rasmussen, J., and Andersen, H. J. 2015. Detecting creeping thistle in sugar beet fields using vegetation indices. Computers and Electronics in Agriculture. 112: pp. 10–19.

Kiani, S., and Kamgar, S. 2011. Application of co-occurrence matrix on wavelet coefficients for crop-weed discrimination. International Journal of Natural and Engineering Sciences. 5(2).

Kim, S., Ryu, C., Kang, Y., and Min, Y. 2015. Improved plant image segmentation method using vegetation indices and automatic thresholds. Journal of Agriculture & Life Science. 49(5): pp. 333–341.

Kincaid, D. T., and Schneider, R. B. 1983. Quantification of leaf shape with a microcomputer and fourier transform. Canadian Journal of Botany. 61(9): pp. 2333–2342.

Kinsman, G. 1993. The history of the lowbush blueberry industry in nova scotia 1950–1990 Nova Scotia Dept. of Agriculture and Marketing.

Kumar, D. A., and Prema, P. 2016. A novel approach for weed classification using Curvelet transform and Tamura texture feature (CTTTF) with RVM classification. International Journal of Applied Engineering Research. 11(3): pp. 1841–1848.

Lamm, R., Slaughter, D., and Giles, D. 2002. Precision weed control system for cotton. Transactions-American Society of Agricultural Engineers. 45(1): pp. 231–248.

Laykin, S., Alchanatis, V., Fallik, E., and Edan, Y. 2002. Image-processing algorithms for tomato classification. Transactions-American Society of Agricultural Engineers. 45(3): pp. 851–858.

Lee, S.H., Chang, Y.L., Chan, C.S. and Remagnino, P. 2016. Plant identification system based on a convolutional neural network for the lifeclef 2016 plant classification task. In Working notes of CLEF 2016 conference.

Lee, D., Schoenberger, R., Archibald, J., and McCollum, S. 2008. Development of a machine vision system for automatic date grading using digital reflective near-infrared imaging. Journal of Food Engineering. 86(3): pp. 388–398.

Lee, W. S., Slaughter, D., and Giles, D. 1999. Robotic weed control system for tomatoes. Precision Agriculture. 1(1): pp. 95–113.

Leemans, V., and Destain, M. 2004. A real-time grading method of apples based on features extracted from defects. Journal of Food Engineering. 61(1): pp. 83–89.

Li, X., and Chen, Z. 2010. Weed identification based on shape features and ant colony optimization algorithm. Paper presented at the International Conference on Computer Application and System Modeling (ICCASM). 2010. V1-384-V1-387.

Liming, X., and Yanchao, Z. 2010. Automated strawberry grading system based on image processing. Computers and Electronics in Agriculture. 71: pp. S32–S39.

Lin, C. 2009. A Support Vector Machine Embedded Weed Identification System. M.S. Thesis, University Illinois, Urbana Champaign, Illinois.

López-García, F., Andreu-García, G., Blasco, J., Aleixos, N., and Valiente, J. 2010. Automatic detection of skin defects in citrus fruits using a multivariate image analysis approach. Computers and Electronics in Agriculture. 71(2): pp. 189–197.

Majumdar, S., and Jayas, D. 2000a. Classification of cereal grains using machine vision: II. color models. Transactions of the ASAE-American Society of Agricultural Engineers. 43(6): pp. 1677–1680.

Majumdar, S., and Jayas, D. 2000b. Classification of cereal grains using machine vision: I. morphology models. Transactions of the ASAE-American Society of Agricultural Engineers. 43(6): pp. 1669–1676.

Majumdar, S., and Jayas, D. 2000c. Classification of cereal grains using machine vision: III. texture models. Transactions of the ASAE. 43(6): pp. 1.

Mao, W., Wang, Y., and Wang, Y. 2003. Real-time detection of between-row weeds using machine vision. ASAE Paper 031004.

Marchant, J., Andersen, H. J., and Onyango, C. 2001. Evaluation of an imaging sensor for detecting vegetation using different waveband combinations. Computers and Electronics in Agriculture. 32(2): pp. 101–117.

Mathanker, S. K., Weckler, P. R., and Taylor, R. K. 2007. Effective spatial resolution for weed detection. Paper presented at the 2007 ASAE Annual Meeting, 1.

Meyer, G. E. 2011. Machine vision identification of plants. *In*: Krezhova, D. (ed.). Recent Trends for Enhancing the Diversity and Quality of Soybean Products. Croatia: InTech.

Meyer, G., Mehta, T., Kocher, M., Mortensen, D., and Samal, A. 1998. Textural imaging and discriminant analysis for distinguishing weeds for spot spraying. Transactions of the ASAE-American Society of Agricultural Engineers. 41(4): pp. 1189–1198.

Meyer, G. E., and Neto, J. C. 2008. Verification of color vegetation indices for automated crop imaging applications. Computers and Electronics in Agriculture. 63(2): pp. 282–293.

Michaud, M., Watts, K., Percival, D., and Wilkie, K. 2008. Precision pesticide delivery based on aerial spectral imaging. Canadian Biosystems Engineering. 50(2): pp. 9–15.

Mohan, A., Parveen, F., Kumar, S., Surendran, S., and Varughese, T. A. 2016. Automatic weed detection system and smart herbicide sprayer robot for corn fields. Ijrcct. 5(2): pp. 055–058.

Montalvo, M., Guerrero, J. M., Romeo, J., Emmi, L., Guijarro, M., and Pajares, G. 2013. Automatic expert system for weeds/crops identification in images from maize fields. Expert Systems with Applications. 40(1): pp. 75–82.

Murphy, C., Lindquist, D., Rynning, A. M., Cecil, T., Leavitt, S., and Chang, M. L. 2007. Low-cost stereo vision on an FPGA. Paper presented at the Field-Programmable Custom Computing Machines. 2007. FCCM 2007. 15th Annual IEEE Symposium on. pp. 333–334.

Nakano, K. 1997. Application of neural networks to the color grading of apples. Computers and Electronics in Agriculture. 18(2): pp. 105–116.

Nieuwenhuizen, A., Tang, L., Hofstee, J., Müller, J., and Van Henten, E. 2007. Colour based detection of volunteer potatoes as weeds in sugar beet fields using machine vision. Precision Agriculture. 8(6): pp. 267–278.

Okamoto, H., Murata, T., Kataoka, T., and HATA, S. 2007. Plant classification for weed detection using hyperspectral imaging with wavelet analysis. Weed Biology and Management. 7(1): pp. 31–37.

Otsu, N. 1979. A threshold selection method from gray-level histograms. IEEE Trans. Sys., Man., Cyber. 9(1): pp. 62–66.

Pearson, T. 2009. Hardware-based image processing for high-speed inspection of grains. Computers and Electronics in Agriculture. 69(1): pp. 12–18.

Pearson, T. 2010. High-speed sorting of grains by color and surface texture. Applied Engineering in Agriculture. 26(3): pp. 499–505.

Pearson, T., Moore, D., and Pearson, J. 2012. A machine vision system for high speed sorting of small spots on grains. Journal of Food Measurement and Characterization. 6(1-4): pp. 27–34.

Pearson, T., Wicklow, D., Bean, S., and Brabec, D. 2013. Sorting of fungal-damaged white sorghum. American Journal of Agricultural Science and Technology. 1(3): pp. 93–103.

Pérez, A., López, F., Benlloch, J., and Christensen, S. 2000. Colour and shape analysis techniques for weed detection in cereal fields. Computers and Electronics in Agriculture. 25(3): pp. 197–212.

Puerto, D. A., Gila, D. M. M., García, J. G., and Ortega, J. G. 2015. Sorting olive batches for the milling process using image processing. Sensors. 15(7): pp. 15738–15754.

Ruiz-Ruiz, G., Gómez-Gil, J., and Navas-Gracia, L. 2009. Testing different color spaces based on hue for the environmentally adaptive segmentation algorithm (EASA). Computers and Electronics in Agriculture. 68(1): pp. 88–96.

Sa, I., Ge, Z., Dayoub, F., Upcroft, B., Perez, T. and McCool, C. 2016. Deepfruits: A fruit detection system using deep neural networks. Sensors. 16(8): pp. 1222.

Shearer, S. A., and Holmes, R. 1990. Plant identification using color co-occurrence matrices. Transactions of the ASAE. 33(6): pp. 1237–1244.

Shearer, S., and Jones, P. 1991. Selective application of post-emergence herbicides using photoelectrics. Transactions of the ASAE. 34(4): pp. 1661–1666.

Shearer, S., and Payne, F. 1990. Color and defect sorting of bell peppers using machine vision. Transactions of the ASAE. 33(6): pp. 1245–1250.

Shinde, A. K., and Shukla, M. Y. 2014. Crop detection by machine vision for weed management. International Journal of Advances in Engineering and Technology. 7(3): pp. 818.

Siddiqi, M. H., Sulaiman, S., Faye, I., and Ahmad, I. 2009. A real time specific weed discrimination system using multi-level wavelet decomposition. International Journal of Agriculture and Biology. 11(5): pp. 559–565.

Singh, J. P. 2014. Designing an FPGA synthesizable computer vision algorithm to detect the greening of potatoes. ArXiv Preprint arXiv:1403.1974.

Sladojevic, S., Arsenovic, M., Anderla, A., Culibrk, D. and Stefanovic, D. 2016. Deep Neural Networks Based Recognition of Plant Diseases by Leaf Image Classification. Computational Intelligence and Neuroscience, 2016.

Tahir, A., Neethirajan, S., Jayas, D., Shahin, M., Symons, S., and White, N. 2007. Evaluation of the effect of moisture content on cereal grains by digital image analysis. Food Research International. 40(9): pp. 1140–1145.

Tang, L., Tian, L., and Steward, B. L. 2000. Color image segmentation with genetic algorithm for in-field weed sensing. Transactions of the ASAE-American Society of Agricultural Engineers. 43(4): pp. 1019–1028.

Tang, L., Tian, L., and Steward, B. L. 2003. Classification of broadleaf and grass weeds using gabor wavelets and an artificial neural network. Transactions-American Society of Agricultural Engineers. 46(4): pp. 1247–1254.

Tang, L., Tian, L., Steward, B., and Reid, J. 1999. Texture-based weed classification using gabor wavelets and neural network for real-time selective herbicide applications. Urbana. 51: 61801.

Teixidó, M., Font, D., Pallejà, T., Tresanchez, M., Nogués, M., and Palacín, J. 2012. An embedded real-time red peach detection system based on an OV7670 camera, ARM cortex-M4 processor and 3D look-up tables. Sensors. 12(10): pp. 14129–14143.

Thompson, J., Stafford, J., and Miller, P. 1991. Potential for automatic weed detection and selective herbicide application. Crop Protection. 10(4): pp. 254–259.

Tian, L. 2002. Development of a sensor-based precision herbicide application system. Computers and Electronics in Agriculture. 36(2): pp. 133–149.

Tian, L. F., and Slaughter, D. C. 1998. Environmentally adaptive segmentation algorithm for outdoor image segmentation. Computers and Electronics in Agriculture. 21(3): pp. 153–168.

Tian, L., Slaughter, D., and Norris, R. 2000. Machine vision identification of tomato seedlings for automated weed control. Transactions of ASAE. 40(6): pp. 1761–1768.

Tuceryan, M., and Jain, A.K. 1998. Texture analysis. In Handbook of Pattern Recognition and Computer Vision (second ed.). World Scientific. Singapore.

Unay, D., Gosselin, B., Kleynen, O., Leemans, V., Destain, M., and Debeir, O. 2011. Automatic grading of bi-colored apples by multispectral machine vision. Computers and Electronics in Agriculture. 75(1): pp. 204–212.

VijayaLakshmi, B., and Mohan, V. 2016. Kernel-based PSO and FRVM: An automatic plant leaf type detection using texture, shape, and color features. Computers and Electronics in Agriculture. 125: pp. 99–112.

Wang, B., Brown, D., Gao, Y., and La Salle, J. 2015. MARCH: multiscale-arch-height description for mobile retrieval of leaf images. Information Sciences. 302(1): pp. 132–148.

Woebbecke, D. M., Meyer, G. E., Von Bargen, K., and Mortensen, D. A. 1993. Plant species identification, size, and enumeration using machine vision techniques on near-binary images. Paper presented at the Applications in Optical Science and Engineering. pp. 208–219.

Woebbecke, D., Meyer, G., Von Bargen, K., and Mortensen, D. 1995a. Color indices for weed identification under various soil, residue, and lighting conditions. Transactions of the ASAE-American Society of Agricultural Engineers. 38(1): pp. 259–270.

Woebbecke, D., Meyer, G., Von Bargen, K., and Mortensen, D. 1995b. Shape features for identifying young weeds using image analysis. Transactions of the ASAE-American Society of Agricultural Engineers. 38(1): pp. 271–282.

Xiaobo, Z., Jiewen, Z., and Yanxiao, L. 2007. Apple color grading based on organization feature parameters. Pattern Recognition Letters. 28(15): pp. 2046–2053.

Xiaobo, Z., Jiewen, Z., Yanxiao, L., and Holmes, M. 2010. In-line detection of apple defects using three color cameras system. Computers and Electronics in Agriculture. 70(1): pp. 129–134.

Yalcin, H. and Razavi, S. 2016, July. Plant classification using convolutional neural networks. In Agro-Geoinformatics (Agro-Geoinformatics), 2016 Fifth International Conference on (pp. 1–5). IEEE.

Yang, W., Wang, S., Zhao, X., Zhang, J., and Feng, J. 2015. Greenness identification based on HSV decision tree. Information Processing in Agriculture. 2(3): pp. 149–160.

Yimyam, P., and Clark, A. F. 2016. 3D reconstruction and feature extraction for agricultural produce grading. Paper presented at the Knowledge and Smart Technology (KST), 2016 8th International Conference on. pp. 136–141.

Yu, Z., Cao, Z., Wu, X., Bai, X., Qin, Y., Zhuo, W., Xiao, Y., Zhang, X., and Xue, H. 2013. Automatic image-based detection technology for two critical growth stages of maize: Emergence and three-leaf stage. Agricultural and Forest Meteorology. 174: pp. 65–84.

Zheng, L., Shi, D., and Zhang, J. 2010. Segmentation of green vegetation of crop canopy images based on mean shift and fisher linear discriminant. Pattern Recognition Letters. 31(9): pp. 920–925.

Zheng, L., Zhang, J., and Wang, Q. 2009. Mean-shift-based color segmentation of images containing green vegetation. Computers and Electronics in Agriculture. 65(1): pp. 93–98.

5

From Manual Farming to Automatic and Robotic Based Farming

An Introduction

*Dan Zhang** and *Bin Wei*

1. INTRODUCTION

Human beings rely on the food to survive. How to achieve productive and efficient farming and therefore provide sufficiently food for human beings will always be an indispensable topic. After the development of advanced robotic machines, the automatic/robotic based farming has become a trend in the agricultural arena. Traditionally, there is dependence on man-power for farming in agriculture. The downside of using man-power is that it relies on a large amount of people and has less efficiency. Shifting to automatic machines, can greatly help farmers in the farming field. Automatic and advanced robotic based farming will become a promising trend in the agricultural and farming areas.

In this chapter, we briefly present the current farming machineries in use at the moment and some issues that we face. There may be other farming machineries also that exist and the authors did not cover in this manuscript. There are numerous sources dealing with the robotic based farming topics and issues. Available sources include books (Kondo et al., 2011; Pedersen et al., 2008), journal publications (Guyer et al., 1986; Mohan et al., 2016; Emmi et al., 2014; Tokekar et al., 2016; Nieuwenhuizen et al., 2007; Hague et al., 2006; Primicerio et al., 2012; Henten et al., 2003; Sa et al., 2006), and conference proceedings (Shibusawa et al., 2000; Werner et al., 2012; English et al., 2013),

Department of Mechanical Engineering, Lassonde School of Engineering, York University, Toronto, Ontario, Canada.
* Corresponding author: dzhang99@yorku.ca

etc. Only a few authors are listed here since there are so many. The following section presents some farming methods which are in use at large.

2. MANUAL FARMING

Farmers usually carry their tools and go farming. The downside of this traditional farming is that it consumes a large amount of manpower and also the farming efficiency is not sound. It is expected that the food demand in the next decade will continue to increase and therefore, efficiency farming is critical, for growing population. The manpower based farming cannot keep pace with the increasing food demand. With the development of modern machinery technology, it is quite possible to transform the manpower based farming fashion to the robotic based farming fashion so that the farming efficiency can be greatly improved and also the manpower can be greatly reduced.

3. ROBOTIC BASED FARMING

3.1 Trackers

Trackers are the most widely used machines in the farming industry. It can perform numerous tasks, such as watering, spraying pesticides, spreading seeds, and harvesting. Trackers can be partially considered as robotic since a tracker needs to be driven by a driver unless the tracker is autonomous type. The advantage of using trackers in the farming field is that trackers can do tasks quickly, but they are usually not very good at precision farming.

3.2 Robotic Gripper

The advantage of the robotic gripper is that it can perform precision farming, however, this type of robot, sometimes, is not quite dexterous as

Fig. 1. A robot gripper.

compared to human hands. How to design dexterous robot grippers that resemble human being's hands will become one of the future works.

3.3 Flying Drones

Nowadays, the flying drones are being used in the farming industry. The drones can carry water or seeds and spread them on the farming ground. The good aspect of using drones is that it can achieve quick watering and seed spreading. Furthermore, the drones can be used for spraying pesticides and monitoring, etc. The applications of flying drones not only can be seen in farming industry, but also they are seen in many other areas, such as sports, video shooting, and military. As the drones conduct, for example, watering or pesticides spraying, the weight of the water or the pesticides that the drones hold is changing, so how to control the drones along with the weight changing situation can be a challenging task.

3.4 Indoor Farming

The concept of indoor vertical farming has been introduced recently, and robots have been used in the indoor farming. The advantages of the indoor farming are that it does not heavily rely on weather condition and also it occupies smaller space as compared to the traditional large space farming.

4. ROBOTIC FARMING ISSUES

There are some main issues in the robotic farming industry that we need to address. For example, sensing issues, robot mechanism design issues, and control issues. For the sensing aspect, how to develop a sensing system that can accurately sense the relative position between the robot and the plant is worth exploring. For the robot mechanism design aspect, we need to figure out how to design a dextrous robot (hands) that resembles the human being's hands. With this robot we can have a more precise and efficient farming outcomes. For the control aspect, we need to figure out how to control the robot so that the precise motion can be achieved, or how to develop a control strategy to cope with the fact that when the robot conduct's farming operations, the outside surroundings' effect will affect the motion of the robot, and this effect should be taken into consideration.

Regarding the control aspect, one can use the model reference adaptive control (MRAC) to address the above mentioned outside surroundings' effect issue. The necessity to employ the MRAC method to a robot is that conventional control technique is not able to handle the load changes situation. During the process of robotic mechanism, end-effector takes different weights of loads, usually the joints' output fluctuates along with time, this phenomenon can deteriorate the end-effector's positioning accuracy performance. However, if one employs the MRAC system, the above issue is effectively rectified and load changes impact is effectively addressed, as demonstrated in Figs. 2 and 3. For detailed studies of the above, please refer to (Zhang and Wei, 2016). The MRAC control system that was developed by scholar Horowitz and subsequently extended by other scholars consists of an adaptation mechanism structure and a position feedback loop structure that is able to detect the error among the joint's ideal position and the joint's real position. This error is then served through the integral section of a PID-like control system, after that the position and velocity feedback values are deducted.

Fig. 2. A 2-DOF vegetable gripper.

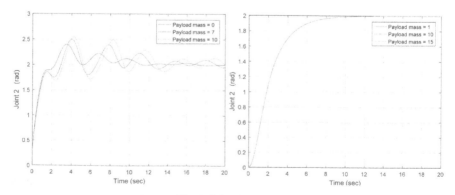

Fig. 3. Joint 2 output.

Based on Fig. 2, and according to the Lagrange technique, the Lagrange of the 2-DOF gripper can be derived as the following,

$$L = K - P$$

$$= \frac{1}{2}(m_1 + m_2)l_1^2 \dot{\theta}_1^2 + \frac{1}{2}m_2 l_2^2 (\dot{\theta}_1 + \dot{\theta}_2)^2 + m_2 l_1 l_2 \cos\theta_2 \dot{\theta}_1 (\dot{\theta}_1 + \dot{\theta}_2) \qquad (1)$$

$$- (m_1 + m_2)gl_1 \sin\theta_1 - m_2 g l_2 \sin(\theta_1 + \theta_2)$$

$$\tau_1 = \frac{d}{dt}\frac{\partial L}{\partial \dot{\theta}_1} - \frac{\partial L}{\partial \theta_1}$$

$$= ((m_1 + m_2)l_1^2 + m_2 l_2^2 + 2m_2 l_1 l_2 \cos\theta_2)\ddot{\theta}_1 + (m_2 l_2^2 + m_2 l_1 l_2 \cos\theta_2)\ddot{\theta}_2$$

$$+ (-2m_2 l_1 l_2 \sin\theta_2)\dot{\theta}_1 \dot{\theta}_2 + (-m_2 l_1 l_2 \sin\theta_2)\dot{\theta}_2^2 + ((m_1 + m_2)l_1 \cos\theta_1 + m_2 l_2 \cos(\theta_1 + \theta_2))g$$

$$\tau_2 = \frac{d}{dt}\frac{\partial L}{\partial \dot{\theta}_2} - \frac{\partial L}{\partial \theta_2} \qquad (2)$$

$$= (m_2 l_2^2 + m_2 l_1 l_2 \cos\theta_2)\ddot{\theta}_1 + (m_2 l_2^2)\ddot{\theta}_2 + (m_2 l_1 l_2 \sin\theta_2)\dot{\theta}_1^2 + m_2 l_2 \cos(\theta_1 + \theta_2)g$$

Applying the PID controller, the output from the controller equals to the torque, therefore,

$$K_p e + K_i \int e\, dt + K_d \dot{e} = \begin{bmatrix} \tau_1 \\ \tau_2 \end{bmatrix} \qquad (3)$$

where error $e = r_p - x_p$. One knows the 2-DOF gripper system's M and N matrices, the accelerations of joint 1 and joint 2 of the gripper are solved by the following ways,

$$\begin{bmatrix} \tau_1 \\ \tau_2 \end{bmatrix} = M\ddot{\theta} + N + Gg$$

$$= \begin{bmatrix} m_{11} & m_{12} \\ m_{12} & m_{22} \end{bmatrix} \begin{bmatrix} \ddot{\theta}_1 \\ \ddot{\theta}_2 \end{bmatrix} + \begin{bmatrix} n_{11} \\ n_{21} \end{bmatrix} + \begin{bmatrix} g_{11} \\ g_{21} \end{bmatrix} g \qquad (4)$$

Thus,

$$K_p e + K_i \int e\, dt + K_d \dot{e} = \begin{bmatrix} \tau_1 \\ \tau_2 \end{bmatrix} = M\ddot{\theta} + N + Gg$$

$$\Rightarrow \begin{bmatrix} \ddot{\theta}_1 \\ \ddot{\theta}_2 \end{bmatrix} = M^{-1}(K_p e + K_i \int e\, dt + K_d \dot{e} - N) \qquad (5)$$

Accelerations of joint 1 and joint 2 are known, a time integral determines the velocities of joint 1 and joint 2, respectively, and second integral determines the positions of joint 1 and joint 2.

$$\begin{bmatrix} \dot{\theta}_1 \\ \dot{\theta}_2 \end{bmatrix} = \int \begin{bmatrix} \ddot{\theta}_1 \\ \ddot{\theta}_2 \end{bmatrix} dt,$$

$$\begin{bmatrix} \theta_1 \\ \theta_2 \end{bmatrix} = \int \begin{bmatrix} \dot{\theta}_1 \\ \dot{\theta}_2 \end{bmatrix} dt \qquad (6)$$

From the MRAC approach, we have the following equation:

$$ControllerOut = \tau = \hat{M}u + \hat{V} - F_p e - F_v \dot{e} \qquad (7)$$

where $u = K_I \int (r_p - x_p) - K_p x_p - K_d x_v$

Since the dynamic formulation for the gripper mechanism is $\tau = Ma + V + Gg$, the accelerations of joint 1 and joint 2 are solved as follows:

$$\hat{M}u + \hat{V} - F_p e - F_v \dot{e} = \tau = Ma + V$$

$$\Rightarrow \begin{bmatrix} a_1 \\ a_2 \end{bmatrix} = \begin{bmatrix} \ddot{\theta}_1 \\ \ddot{\theta}_2 \end{bmatrix} = M^{-1}(\hat{M}u + \hat{V} - F_p e - F_v \dot{e} - V) \qquad (8)$$

Furthermore, the adaptive algorithm can be determined as the following,

$$\int_0^T y^T(t) \tilde{M} u(t) dt = \int_0^T \begin{bmatrix} y_1 \\ y_2 \end{bmatrix}^T \begin{bmatrix} \tilde{m}_{11} & \tilde{m}_{12} \\ \tilde{m}_{12} & \tilde{m}_{22} \end{bmatrix} \begin{bmatrix} u_1 \\ u_2 \end{bmatrix} dt$$

$$= \int_0^T [y_1, y_2] \begin{bmatrix} \tilde{m}_{11} & \tilde{m}_{12} \\ \tilde{m}_{12} & \tilde{m}_{22} \end{bmatrix} \begin{bmatrix} u_1 \\ u_2 \end{bmatrix} dt \qquad (9)$$

$$= \int_0^T \tilde{m}_{11} y_1 u_1 dt + \int_0^T \tilde{m}_{12} (y_1 u_2 + y_2 u_1) dt + \int_0^T \tilde{m}_{22} y_2 u_2 dt$$

By observing the first term in equation (9), one needs to satisfy $\frac{d}{dt} \hat{m}_{11}(t) = \frac{d}{dt} \tilde{m}_{11}(t)$, so that $\int_0^T \tilde{m}_{11} y_1 u_1 dt \geq -\gamma^2$.

Since

$$\int_0^T z(t)^T \dot{z}(t)\,dt = \frac{z(T)^T z(T)}{2} - \frac{z(0)^T z(0)}{2} \geq -\frac{z(0)^T z(0)}{2} = -\gamma_0^2$$

So by selecting

$$\frac{d}{dt}\hat{m}_{11}(t) = \frac{d}{dt}\tilde{m}_{11}(t) = k_{m11}y_1 u_1,$$

$$\Rightarrow y_1 u_1 = \frac{\dot{\tilde{m}}_{11}(t)}{k_{m11}} \tag{10}$$

Thus,

$$\int_0^T \tilde{m}_{11}y_1 u_1\,dt = \int_0^T \tilde{m}_{11}\,\dot{\tilde{m}}_{11}\,\frac{1}{k_{m11}}\,dt = \frac{1}{k_{m11}}\int_0^T \tilde{m}_{11}\,\dot{\tilde{m}}_{11}\,dt \geq -\gamma^2$$

By employing the same approach and applying to the other two terms, the following equation is obtained,

$$\frac{d}{dt}\hat{m}_{12}(t) = \frac{d}{dt}\tilde{m}_{12}(t) = k_{m12}(y_1 u_2 + y_2 u_1),$$

$$\frac{d}{dt}\hat{m}_{22}(t) = \frac{d}{dt}\tilde{m}_{22}(t) = k_{m22}y_2 u_2 \tag{11}$$

Similarly, we can determine the adaptive algorithm for N as the following,

$$\frac{d}{dt}\hat{n}_{12}(t) = \frac{d}{dt}\tilde{n}_{12}(t) = k_{n12}(2y_1 x_{v1} x_{v2} - y_2 x_{v1}^2),$$

$$\frac{d}{dt}\hat{n}_{22}(t) = \frac{d}{dt}\tilde{n}_{22}(t) = k_{n22}y_1 x_{v2}^2 \tag{12}$$

In (Sharifi et al., 2014), a model reference based adaptive impedance controller was designed by combining the MRAC and impedance control for tracking control problem in human–robot interaction. In (Huh and Bien, 2007), a sliding mode based MRAC was proposed for a robotic manipulator. By introducing the sliding model control approach to the MRAC, the control system allows the manipulator to follow its nominal dynamics. In (Kamalasadan and Ghandakly, 2008), a fuzzy multiple-reference-model generator-based MRAC scheme was proposed by combining a fuzzy logic switching strategy and a direct MRAC algorithm. In (Su, 2007), a model

Fig. 4. Dexterous robot hand design.

reference was designed by inserting a PID controller to the feedback path for robot motion control. In (Suboh et al., 2009), a fuzzy MRAC was proposed by introducing the Takagi-Sugeno-fuzzification control and P-I control to the MRAC. Based on the studies in (Sharifi et al., 2014; Huh and Bien, 2007; Kamalasadan and Ghandakly, 2008; Su, 2007; Suboh et al., 2009), approaches for new controller design for robotic manipulators can be achieved by combining MRAC and other control system to design advanced MRAC system, so as to cope with the above mentioned control issues better and effectively in the farming industry.

Regarding the robot mechanism design aspect, as shown in Fig. 4, the robot hand on the left side of the figure is very rigid and is not as dexterous as human hands, as shown on the right side of the figure. The next step would be is to design robotic hands that resemble human hands, in order to make the farming more precise. Another issue would be how to control the robotic hand to make its motion more adaptive, i.e., make it move like human hands. Learning control approach is one of the strategies that we can consider, so as to make the control system more intelligent.

5. FINAL REMARKS

In this paper, some of the main robotic based machineries were presented, that are used in farming, i.e., trackers, robotic grippers, flying drones, and indoor farming. The robotic and automatic based farming will replace the manpower based farming in the near future, and the robots will become the main labour force in the agriculture and its applications. Some issues for the robotic farming are also briefly addressed. This paper can provide a general guideline for future research in the field of robotic based farming.

6. ACKNOWLEDGEMENTS

The authors would like to thank the financial support from the Natural Sciences and Engineering Research Council of Canada (NSERC). The authors gratefully acknowledge the financial support from the Canada Research Chairs program.

REFERENCES

Emmi, L., Gonzalez, M., Pajares, G., and Gonzalez, P. 2014. New trends in robotics for agriculture: integration and assessment of a real fleet of robots. The Scientific World Journal. pp. 1–21.

English, A., Ball, D., Ross, P., Upcroft, B., and Wyeth, G. 2013. Low cost localisation for agricultural robotics. Proceedings of the 2013 Australasian Conference on Robotics & Automation. pp. 1–8.

Guyer, D., Miles, G., Schreiber, M., Mitchell, O., and Vanderbilt, V. 1986. Machine vision and image processing for plant identification. Transactions of the ASAE. 29(6): pp. 1500–1507.

Hague, T., Tillett, N., and Wheeler, H. 2006. Automated crop and weed monitoring in widely spaced cereals. Precision Agriculture. 7(1): pp. 21–32.

Henten, E., Tuijl, B., Hemming, J., Kornet, J., Bontsema, J., and Os, E. 2003. Field test of an autonomous cucumber picking robot. Biosystems Engineering. 86(3): pp. 305–313.

Huh, S., and Bien, Z. 2007. Robust sliding mode control of a robot manipulator based on variable structure-model reference adaptive control approach. IET Control Theory & Applications. 1(5).

Kamalasadan, S., and Ghandakly, A. 2008. A novel multiple reference model adaptive control approach for multi modal and dynamic systems. Control and Intelligent Systems. 36(2): pp. 119–128.

Kondo, N., Monta, M., and Noguchi, N. 2011. Agricultural Robots: Mechanisms and Practice. Trans Pacific Press.

Mohan, A., Parveen, F., Kumar, S., Surendran, S., and Varughese, T. 2016. Automatic weed detection system and smart herbicide sprayer robot for corn fields. International Journal of research in Computer and Communication Technology. 5(2): pp. 55–58.

Nieuwenhuizen, A., Tang, L., Hofstee, J., Müller, J., and Henten, E. 2007. Colour based detection of volunteer potatoes as weeds in sugar beet fields using machine vision. Precision Agriculture. 8(6): 267–278.

Pedersen, S. M., Fountas, S., and Blackmore, S. 2008. Agricultural Robots—Applications and Economic Perspectives, Service Robot Applications, Yoshihiko Takahashi (ed.). ISBN: 978-953-7619-00-8.

Primicerio, J., Gennaro, S., Fiorillo, E., Genesio, L., Lugato, E., Matese, A., and Vaccari, F. 2012. A flexible unmanned aerial vehicle for precision agriculture. Precision Agriculture. 13(4): pp. 517–523.

Sa, I., Ge, Z., Dayoub, F., Upcroft, B., Perez, T. and McCool, C. 2006. Deep fruits: A fruit detection system using deep neural networks. Sensors. 16(8).

Sharifi, M., Behzadipour, S., and Vossoughi, G. 2014. Model reference adaptive impedance control in Cartesian coordinates for physical human–robot interaction. Advanced Robotics. 28(19): pp. 1277–1290.

Shibusawa, S., Monta, M., and Murase, H. 2000. Bio-robotics, Information Technology, and Intelligent Control for Bioproduction Systems. Proceedings of 2nd IFAC/CIGR International Workshop, Sakai, Osaka, Japan, 25–26 November, 2000.

Su, W. 2007. A model reference-based adaptive PID controller for robot motion control of not explicitly known systems. International Journal of Intelligent Control and Systems. 12(3): pp. 237–244.

Suboh, S., Rahman, I., Arshad, M., and Mahyuddin, M. 2009. Modeling and control of 2-DOF underwater planar manipulator. Indian Journal of Marine Sciences. 38(3): pp. 365–371.

Tokekar, P., Hook, J., Mulla, D., and Isler, V. 2016. Sensor planning for a symbiotic uav and ugv system for precision agriculture. IEEE Transactions on Robotics. 32(6): pp. 1498–1511.

Werner, R., Muller, S., and Kormann, K. 2012. Path tracking control of tractors and steerable towed implements based on kinematic and dynamic modeling. In 11th International Conference on Precision Agriculture, Indianapolis, Indiana, USA. pp. 15–18.

Zhang, D., and Wei, B. 2016. Study on Payload Effects on the Joint Motion Accuracy of Serial Mechanical Mechanisms, Machines. 4(4): pp. 21; doi:10.3390/machines4040021.

6

Cooperative Robotic Systems in Agriculture

Khalid Salah* and XiaoQi Chen

1. INTRODUCTION

The availability of skilled workers in agricultural areas such as greenhouses, orchards, plantations, and forestry are declining. Further, difficult working conditions in agricultural environments, has encouraged the automation of some tasks. Besides, current commercial agricultural practices are standard and systematic which could be easily represented into automation algorithms. A few crops which are being massively produced and have large plantation's volume have been already automated such as cotton, corn, or wheat. Some agricultural activities require accurate and robust system due to their complex environments and challenging conditions like orchards or greenhouses (Wang et al., 2016). Fruits transportation during harvesting process in a commercial orchard is a feasible application to be automated in order to optimize the overall cost and harvesting time. A team of robotic agents (RAs) can collaboratively assist pickers and transport collected fruit bins to a loading station.

Coupled with automated systems' sensible solutions, RAs should have intelligent tools in order to enable a robust response to new tasks with dynamic conditions (Barth et al., 2014). RAs can provide advanced controllability and flexible kinematics to overcome such complexity (Wang et al., 2016). RAs have been widely used to automate agricultural operations with the intention to overcome some challenging aspects such as dynamic environment, nonlinearity, complicated modeling, safety or collision avoidance, formation and configuration, technology limitation,

The University of Canterbury, Department of Mechanical Engineering, Christchurch, New Zealand.
Email: xiaoqi.chen@canterbury.ac.nz
* Corresponding author: khalid.salah@pg.canterbury.ac.nz

overall cost, and being user-friendly (Li et al., 2015). Series of actions must be considered when applying fully autonomous systems to ensure effectiveness, safe production, good localization, obstacle and point of interest detection, and communication (Emmi et al., 2014).

Besides RAs' proven capability, RAs have been given advantages of advanced sensors and actuators in order to be equipped with suitable agricultural tools. The sensing technologies developed in the last two decades have allowed accurate positioning and reliable performance. Utilizing multiple advanced electronics in robots not only improve RAs reliability but also increase their overall cost and complexity. However, previous studies concluded that simple hardware designs for RAs in agricultural minimize total cost and system's complexity. Thus, simple RAs can be upgraded with an integrated implement which is an actuator to perform a certain task such as spraying, weeds removal, fertilizing, and seed planting (Emmi et al., 2014).

Granted that an advanced RA has better performance and is more robust than a simple RA, however, researchers have ratiocinated that a cooperative team of simple RAs has better accuracy in localization, navigation, path planning, and optimal performance. The combination of two or more interacting intelligent agents is referred to as a multi-agent system. A multi-agent system is a swarm intelligence system having a smart team of agents effectively interacting with each other to complete common tasks. Multi-agent systems have the ability to resolve complicated tasks which are difficult or impossible for a single agent to accomplish (Barca and Sekercioglu, 2013). Multi-agent systems in cooperative environments allow more sophisticated agents to share their capabilities with other agents which have limited capabilities (Bailey et al., 2011). Multi-agent systems have the robustness of a single agent; thus, they have been applied successfully to agriculture and manufacturing applications.

An equally important aspect to be considered when applying a multi-agent system is its control architecture. The selection of a suitable control topology is an essential part of the multi-agent system. It can be categorized in two topologies which are centralized control and decentralized control. Each of these control strategies have several advantages and disadvantages. A hybrid system would combine both schemes' advantages and overcome weaknesses (Barca and Sekercioglu, 2013). Multi-agent's control structure within agriculture applications should allow its agents to be cognitive in order to maintain robustness, manage trajectories and predict other agents' trajectories which are achieved simultaneously by updating and calculating the spatiotemporal trajectory of every collaborative agent in the fleet. The system's control manages trajectory data mining which is defined as deriving, pre-processing, uncertainty managing via map-matching, pattern mining, classifying, anomalies detecting, and transferring data to another representation (Zheng, 2015).

Due to the complex data processing and background computations performed by the controller, robotic systems should have operating systems which can integrate multiple hardware and software modules. Besides their design complexity, a robotic system should be user-friendly with flexible programming interface and more crucially have a flexible middleware which can be customisable to different situations and applications. A middleware should meet design diversities and also be compatible with multiple sensors and actuators of different designs and manufactures, to process data and execute commands. It should be flexible to handle applications' development since robots are made of heterogeneous components. They are also required to interact with different communications and processing mechanisms instruments, integration with other systems such as agricultural implements, existing software libraries or algorithms and have the ability to collaborate or share information with other systems. Previously developed middlewares are, Ocra, UPnP, RT-Middleware, ASEBA, Player/Stage, PEIS Kernel, ORiN, MARIE, RSCA, MARIE, Middleware of AWARE, Sensory Data Processing Middleware, Distributed Humanoid Robots Middleware, Layer for Incorporation, WURDE, OROCOS, and ROS which is a recent and widely used framework (Mohamed et al., 2008). ROS also is an open-sourced framework which is compatible with multiprogramming languages and provides standard operating functions such as hardware perception, low-level actuators' control, coding, and implemented operations, message and command communication between nodes, and packages' management (Wang et al., 2016).

The need for more sophisticated systems in industries led to the emergence of multi-agent control systems (Rodrigues et al., 2013). Given these points, studies on swarm intelligence and multi-agent systems have received significant attention recently, however, many aspects and research areas remain to be explored (Gautam and Mohan, 2012). The advantages of applying multi-agent systems in agriculture are propitious. Such system can redistribute complicated agricultural tasks into smaller and more practical parts for optimal performance (Pentjuss et al., 2011). The distribution of multiple operating agents could maintain a more accurate operation since an agent's failure can be compensated by another available agent. A fault tolerant and a goal oriented system, utilizing a multi-agent system can increase agriculture production effectively (Pitla et al., 2010).

In this chapter, multi-agent systems in agricultural applications involving a RA to RA and RA to human agent (HA) collaboration are reviewed. Common systems' control architecture and design, tools and middleware, planning and decision execution, cooperation behaviour, and communication systems are discussed for recently developed systems for agricultural applications. A case study of a multi-agent collaborative system's framework for transporting harvested fruit bins in orchards, which is investigated at the University of Canterbury, is presented.

2. SYSTEMS' ARCHITECTURE AND DESIGN

Having a single robot operating in an open space application, such as agriculture, requires human supervision in order to monitor its performance. Also, current legislations of many countries do not allow full autonomous machinery without direct or indirect human's supervision. In such cases, a HA is required for each operating robot. Supervising each single RA increases the total cost and negates the need for automation. Utilizing a multi-agent system, with only one HA observing the whole fleet (Noguchi et al., 2004) would resolve the supervising issue with an optimal overall cost. The design of cooperative multi-agent systems is application- and functionality-based and include factors such as control structure, middleware, and navigation accuracy.

2.1 Control Topology

The structure of a cooperative multi-agent system control and its features reflect its capability, level of cooperation, limitation, and ability to expand the number of its participating agents. Most widely used control architectures among cooperative systems are decentralized control and hybrid control system since they have the advantages of being more robust than a centralized control system (Cao et al., 1997).

2.1.1 Centralized control

A centralized control topology has an advanced robot or a computer with a powerful processor as a leader to plan for the entire fleet, applying series of algorithms to perform specific actions. The central controller is in charge of collecting and processing data of each individual agent to plan tasks' execution on a global integrated level based on a prior knowledge of each agent's current status (Garro et al., 2007). A centralized control system has the advantage of producing optimal plans by maintaining data communication and feedback from the entire system, directly controlling each agent, and thus can predict behaviour and result of the system. However, there are some disadvantages when applying centralized control systems such as, achieving an optimal communication coverage, the computational cost of larger fleets, and total failure during the absence or failure of the leader. In addition, it is totally dependent on the leader's direct communication range, which results in a temporary failure if leader is out of range (Dias and Stentz, 2000).

Most common multi-agent agricultural centralized control architectures are in the form of a master to slave topology. This topology allows a sophisticated agent or computer to function as a master and the other robots as slaves. A master to slave multi-robot control structure was developed by Noguchi et al. (2004) for farming operations. It was powered by the GOTO algorithm which was developed as a motion algorithm to allow a slave robot to move from its current point to another point planned by the master. Another algorithm which was also developed is called the FOLLOW algorithm allowing a slave robot to mimic the master's navigation with an offset distance as shown in Fig. 1. Both algorithms considered pathway planning, collision detection, and avoidance, besides speed and steering control. The proposed system can be used for harvesting and transporting hay or corn. The GOTO algorithm was computer simulated and a risk index was maintained in order to avoid a master and slave collision. The closest distance between the master and slave robot while decreasing speed of the slave robot is 12.5 m with a risk index of 0.46. Another method to avoid master and slave collision is to alter the slave's path and a 12.6 m distance was achieved with a safety

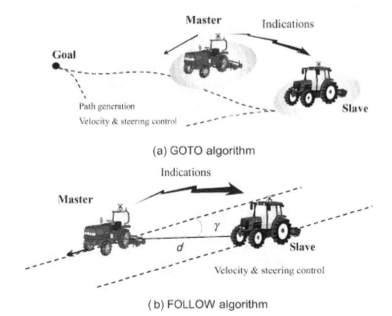

(a) GOTO algorithm

(b) FOLLOW algorithm

Fig. 1. Centralized control system and the proposed GOTO and FOLLOW algorithms for a master–slave robot system (Noguchi et al., 2004).

index of 0.46. The FOLLOW algorithms has an overshoot of 0.134 m and 0.184 m RMS error when applying the sliding mode controller and PD controller respectively. On the other hand, the sliding mode controller achieved a RMS error of 0.106 m than 0.131 m for PD controller.

2.1.2 Decentralized control system

A decentralized or distributed control system is a control topology allowing each agent to operate on its local information to achieve a common task. It enhances agents' autonomy by processing sensors' data, maintaining tasks planning, managing communication coverage, and executing commands independently. This system is preferred over a centralized control system since it is more robust to central failure, has less communication coverage limitation within the system, can accommodate a larger number of agents, and resolves difficulties of the multi-agent tasks coordination problem. A distribution control not only overcomes central control's disadvantages, but also enables the system to break a complicated task into sub-tasks allowing continuous and fast response to dynamic conditions and enhanced collaborative behavior (Fidan et al., 2007). However decentralization would result in output oscillation and wastage of power due to the absence of central tracking to ensure stability (Ma and Yang, 2005).

Considering the advantages, a smartweed treatment heterogeneous multi-agent system was designed (Kazmi et al., 2011) to investigate technological challenges of guiding and estimating a heterogeneous multi-agent system. A decentralized control structure was adapted to control two unmanned aircraft systems (UAS) and an unmanned ground vehicle (UGV) equipped with advanced vision sensors as shown in Fig. 2. A Weed detection process was achieved by processing images obtained by a Multispectral camera and Time-of-Flight camera. Its data exchange process allowed each agent to evaluate the overall task and handle their sub-tasks individually. The communication range limitation affected the data exchange process among the fleet, since UAS has a wider but distanced observation while UGV has a closer but narrow inspection. It

Fig. 2. Heterogeneous and autonomous agents (from left): Vario XLC, Maxi Joker-3 and robuROC-4 by (Kazmi et al., 2011).

was concluded that having heterogeneous multi-agent is more complex but has a better flexibility towards wider ranges of applications and more customized solutions.

2.1.3 Hybrid control systems

A hybrid control system is a result of integrating a centralized control with a decentralized control in a hierarchical structure to get advantages of a decentralized control flexibility along with a high-level control which have the ability to plan tasks and monitor the performance of the participating agents (Barca and Sekercioglu, 2013). A complex in-memory distributed computation involving very large data sets generated by each agent in a fleet can be stored on hard drives or larger memory computer to provide locality-aware scheduling, fault tolerance, recovery from failures, and load balancing. An advanced and powerful processor robot or computer collects data and keeps track of each agent while each agent which is decentralized control will have the awareness to manage individual tasks locally. The hybrid system can overcome the complication of pure centralized and decentralized structure; thus, it is a practical design for complicated multi-agent operations (Cheng et al., 2008).

Emmi et al. (2014) developed a fully hybrid integrated control system architecture for individual robot and fleet of robots working together. It was designed by integrating autonomous vehicles and autonomous implements, which are devices carried or pulled by a vehicle to perform a certain function such as herbicide, pesticide booms, mechanical or thermal weed removal which have separate controllers and can be controlled externally. The vehicle used is CNH Boomer-3050 which was modified and equipped with a Weed Detection System, a crop row detection system, a laser range finder to detect obstacles, communication equipment, a two-antenna global positioning system, an Inertial Measurement Unit (IMU), a vehicle controller which is in charge of computing steering control laws, throttle, braking for path tracking purposes, a central controller as a decision making system and fuel cell as an additional energy power supply. The Robot Fleets for Highly Effective Agriculture and Forestry Management (RHEA) topology consists of an external computer in a base station, user portable device to allow human supervision, wireless communication medium, and a fleet of mobile units as shown in Fig. 3. The integration of ground mobile unit controller (GMU) with the main controller, improved reaction capabilities to speed change and trajectories which were continuously evaluated and improved. The system was very efficient, easily integrated to new hardware and sensors, had sophisticated algorithms, and allowed full autonomy and better collaboration. The

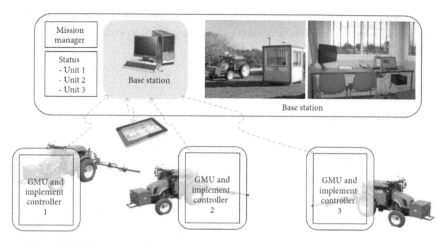

Fig. 3. Hybrid control with a central base station and fleet autonomous vehicle with autonomous implements (Emmi et al., 2014).

proposed topology successfully minimized RHEA's hardware and improved the processing time since image acquisition, image processing, and image sharing took 80–160 ms, 200–250 ms, and 1 ms respectively which was faster than the previous RHEA design. Additionally, these processing times were obtained while another four processes were being executed in parallel. However, the system is complex and has expensive instruments.

2.2 Middleware and Tools

Middleware is a user-friendly programming interface linking high level controller with operating low-level actuators. It supports integrating hardware and software modules efficiently. It is crucial for robotic middleware to have unique characteristics which would enable robust robotic applications, adapt to different scenarios, meet designs diversity, and enhance applications' developments. Mohamed et al. (2008) and Min Yang et al. (2010) individually investigated common robotic frameworks developed for robotics application. Both studies concluded that robotic middleware should have flexible architecture and characteristics to deliver customized solutions in order to develop required applications.

The most widely used open source framework defined as Robot Operating System (ROS), is a collection of software frameworks for robot software development, providing operating system-like functionality on a heterogeneous computer cluster (Wang et al., 2016). ROS is also currently preferred by robotics developers since it can manage command execution in Python or C++ language as messages in parallel through assigned nodes.

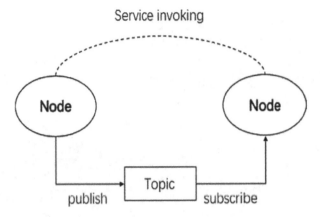

Fig. 4. Exchanging message of a topic in parallel between publisher and subscriber (Wang et al., 2016).

In addition, ROS's communication medium is either Ethernet or Wireless based on a common method such as publish-subscribe or event-driven communication to relevant generated data called topic. In these methods, the nodes communicate data continuously without earlier knowledge who communicates with whom as shown in Fig. 4. Another communication method based on a request which ROS software uses is called request-reply communication which is support via services. The communication methods known as actions are used when a node is required to monitor or supervise certain actions. The node continuously gets feedback, therefore, can cancel or redirect an action.

The most compelling evidence is that ROS works to a satisfactory level by supplying useful tools which enable inspection, visualization, debugging, mapping and localization, and integration such as Rviz and Gazebo with other open-source libraries such as OpenCV, PCL, and MoveIt. ROS has standard message's formats with stable publishing frequency and accepts customized messages that publishers and subscribers agreed to. ROS user communities are very active, thus, solutions to different problems can be founded in ROS Wiki or online. ROS 3D visualization tool which is called Gazebo enables design visualization and helps in developing the platform, running algorithms, software testing and building modules. It also supports simulating sensors and actuators. Moreover, virtual sensors and actuators in Gazebo generate data which are similar to the real world generated data from actual sensors and actuators. ROS enables offline optimization since it can store all the generated data with a time stamp (Linz et al., 2014). As a result (Barth et al., 2014), investigated the experience of using ROS as a middleware for developing an agricultural robot. Technical aspects which were discussed

Fig. 5. Dual-arm tomatoes harvester in farm and the 3D simulation using Rviz Tool (Barth et al., 2014).

are, sensing perception, manipulators, system framework and mission control. The study examined ROS's design capability and methods by designing a tomatoes harvesting dual-arm robot which was tested in a real farming environment. The system's 3D model was built and visualized by Rviz tool and a motion planning process was supported by Moveit library which is an integrated toolkit in ROS as shown in Fig. 5.

ROS was found to have several disadvantages such as, learning ROS takes considerable time, newly developed versions of ROS lack compatibility with older versions, and older versions need to be customized with the newer release. The debugging process can be challenging since communication is handled through messages; thus, they require isolating and simulating each message. ROS does not support real-time response to other external softwares even if they are linked to it. It was however, concluded that ROS has wider and applicable roles in robotics' development in the future.

2.3 Collaborative Navigation

It is important for a RA to be able to navigate, build collision-free paths and find its next step such as where to go or what to do. A RA should manage its navigation's subtasks which are self-localization, path planning, maps building, and map utilization irrespective of whether a robotic agent is located in an indoor or outdoor environment. Therefore, indoor or outdoor navigation can be generally categorized as completely known, partially known, and unknown. Current sensing technologies and powerful programming enhance locating a goal point, path planning,

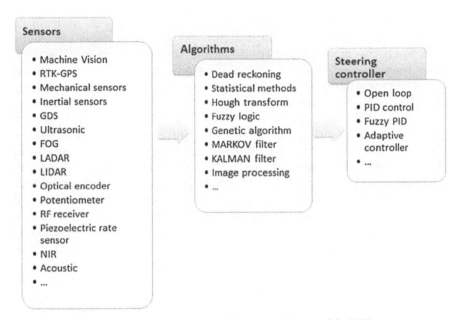

Fig. 6. Basic navigation control diagram (Mousazadeh, 2013).

and avoiding encountered obstacles (Khan and Ahmmed, 2016). Each navigation subtask can be categorized as deterministic or reactive, for example path tracking is deterministic while avoiding obstacles is reactive (Vougioukas et al., 2005). A RA's movement in agricultural applications is straightforward due to the standard row plantation patterns either in outdoor farms and orchards or indoor plantations and greenhouses.

Accordingly, Mousazadeh (2013) reviewed the navigation process of autonomous agricultural vehicles and compared different navigation algorithms in terms of accuracy and speed. In the review, navigation systems were characterized in six categories which are, dead reckoning, image processing, statistically based developed algorithms, fuzzy logic control, neural network, genetic algorithm, and Kalman filter based algorithms. The current autonomous systems use sensors' collected data to be fed to their navigation algorithms. The navigation algorithm plans and executes next step such as, moving, steering or stopping. The most commonly used sensors in agricultural navigation are vision based, Real Time Kinematic-Global Positioning System (RTK-GPS), mechanical sensor, inertial sensors, Geomagnetic Direction Sensors (GDS), ultrasonic, Fiber-optic Gyroscope (FOG), Laser Radar (LADAR), Light Detection and Ranging (LIDAR), optical encoder, Radio Frequency receiver (RF receiver), piezoelectric Yaw-rate sensor, Near Infra-Red (NIR), and Acoustic sensor as shown in Fig. 6.

In addition, Keicher and Seufert (2000) analysed agricultural autonomous vehicles and implements navigation systems in Europe and listed the most widely used sensors such as mechanical sensors, laser triangulation, machine vision, ultrasonic, geomagnetic, and global navigation satellite (GPS) systems, which generate position, altitude, and direction of robots' movement.

Sharifi et al. (2016) developed and tested a Mobile Autonomous Robot for Intelligent Operations (MARIO) at the University of Canterbury, New Zealand. The MARIO's navigation system integrated visual odometry (VO) and inertial measurement unit (IMU) and fused their generated data to an Extended Kalman Filter and VO algorithms in order to self-localize the robot in a GPS-denied environment. A comparison of using two open source algorithms which are called *Favis* and *Libviso* was presented. It was concluded that *Libviso* achieved a better accuracy than *Favis*. MARIO was developed on a Robotic Operating System (ROS) middleware which enabled simulating sensors' data and visualizing its experimental testing in Gazebo and Rviz tools (Sharifi et al., 2016).

Collaborative multi-agent system's navigation utilizes each agent's navigation ability. In other words, it is the combination of navigation subtask of each single agent and it is also dependent on the control system architecture. There are two common scenarios in collaborative navigation, the first scenario is that all agents navigate to a single location while in the second scenario each agent navigates to a different targeted location with the aid of other agents' presence (Bayındır, 2016). The second scenario is more practical in agricultural applications since autonomous agents either with or without an implement are designed for a specific task, without causing damage to plantation and soil. Thus, an iterative process requires a previous knowledge of a route from the agent's current location to the desired target and also the knowledge of other agents' poses.

Likewise, a multi-agent navigation process is more effective than single-agent system due to the information exchange mechanism among robots which establishes a previous knowledge of the surrounding environment. A good communication coverage with appropriate algorithms and consecutive data exchange between agents are essential for accurate navigation. Collaborative navigation routing algorithms can be categorized as static routing or dynamic routing. A static routing allows an agent to follow a sequence of fixed landmarks while dynamic routing depends on direct communication with another neighboring agent to determine a targeted location from a current location (Wurr and Anderson, 2004).

Notably, the hybrid control multi-agent application known as Robot Fleets for Highly Effective Agriculture and Forestry Management (RHEA) implemented three levels of navigation subsystems. Level one which is a combination of navigation sensors namely RTK-DGPS, machine vision, LIDAR, and IMU. Level two is a navigation planner performing paths' generation and obstacle avoidance. Level three is the navigation execution layer which has path following, steering, and throttle controllers (Emmi et al., 2014).

Adamides et al. (2012) improved a HA to a RA collaborative system to execute spraying tasks and navigation by introducing a semi-automatic teleoperation. The design's principles such as visibility, safety, simplicity, feedback, extensibility, and cognitive load reduction were introduced to allow a human agent to contribute to robot navigation, target selection, and spraying. Vougioukas et al. (2012) developed a multi-agent robotic system to transport bins of harvested fruits from a fixed position to a drop station. The Split Delivery Vehicle Routing Problem (SDVRP) which is a navigation and formation algorithm was used to route different size robotic agents to bin's location and then to a drop station and it was developed using the Mixed Integer programming method.

Furthermore, English et al. (2013) investigated and developed a robust pose estimation method to common sensors failures by combining multiple low-cost sensors on small and light robotic farming machinery. The used sensors are low-cost GPS, inertial sensors, and vision-based row tracking. The integration of GPS and inertial sensors with the vision-based row tracking sensor enabled the system to overcome long signals interruptions and repeated dropouts. It is also a lower cost option than expensive GNSS navigation systems. The added vision system allows robots to observe the visual features while driving in a way that mimics humans' driving and following GPS's directions as shown on the localization block diagram Fig. 7. Physical experiments were carried out by a robot which covered 6 hectares and resulted in 0.18 m root means square (RMS) pass-to-pass errors while 95th percentile error was 0.28 m and all errors were less than 0.5 m wide side-spray. The missed out area was 2.6% and the repeatedly covered area was 9.7%. A correction on the IMU constant bias reduced RMS errors by 28% and percentile errors by 42%.

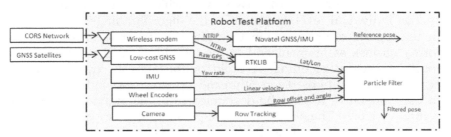

Fig. 7. Localization system components (English et al., 2013).

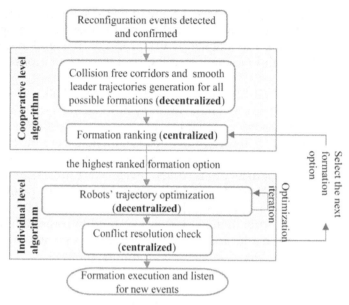

Fig. 8. Algorithm architecture (Li et al., 2015).

A HA to a RA cooperative agricultural system was also developed by Farangis Khosro et al. (2014) to aid a picker during a strawberry harvesting process by transporting filled tray from a picker location to a loading station. The RA also monitored picker's posture since long and continuous bending during picking may result in a low back disorder (LBD). The designed platform allowed navigation in narrow strawberry furrows and it was equipped with ultrasonic sensors mounted in the front and back allowing collision avoidance, straight motion, and safe movement which was also feeding data to an Arduino microcontroller executing navigation and furrow path stabilization algorithms.

Along with the above studies, Li et al. (2015) studied a hierarchical decision method and trajectory planning for a group of collaborative agricultural robots performing tasks such as citrus harvesting. A framework algorithm handled two levels which are a cooperative level and an individual level as shown in the algorithm architecture Fig. 8. In this algorithm optimization took place at the cooperative level for formation task assignment and at the individual level for agricultural robots' trajectory planning. A rapid optimization of trajectories was achieved by the proposed algorithm and a performance index was added to the cooperative level to be decoupled from the individual level and control variables. The adoption of a re-planning strategy enabled robots to adapt to dynamically changing environments.

3. ROBOTIC COOPERATIVE BEHAVIOUR

A multi-agent system cooperative behaviour is determined based on its communication ability, amount and type of exchanged data, designs' similarities, and common tasks to be achieved. An artificial communication can be categorized into two types. The first type is via observation similar to biological stigmergy which is communicating through observing surrounding changes or signs (Holland and Melhuish, 1999). The second type is via messages which are passing packets of data containing specific information such as machine IDs, pose, time stamp, velocity, and the status of an assigned task. The messages' communication is managed via Bluetooth, wireless, infrared, or 3G GSM internet depending on the team size and communication range. The type of information shared between agents specifies the degree of collaboration between them, for example, the simple exchanged information might contain only pose and time stamp and more complex information may contain commands, direction instructions, a request of a specific agent to do a specific task, or algorithms update (Capodieci and Cabri, 2013).

A complicated inter-robot communication network has two basic concepts which are an implicit communication and explicit communication. The implicit communication is achieved when an agent broadcasts its status data to the whole fleet while an explicit communication is a point to point communication with a specific agent. Whether to use an implicit, explicit communication or both combined depends on the assigned task. The cooperation behaviour was divided into three categories such as no cooperation, modest cooperation, and absolute cooperation by Pitla et al. (2010). They investigated a multi-robot system control architecture (MRSCA) for agricultural production. The MRSCA tested the three cooperative behaviour categories supporting different tasks efficiently.

3.1 No Cooperative Behaviour

A multi-agent system would have no cooperative behaviour if the system or its agents have implicit communication only. The system remains collaborative but has no or less cooperation behaviour. In other words, agents in the system still performed assigned tasks collaboratively, but do not have a point to point communication, nevertheless, they still broadcast their status. Most of the homogenous multi-agent applications in agricultural applications such as spraying, planting, and fertilizer require implicit communication only (Pitla et al., 2010).

A safe collaborative navigation system in an orchard or in a farm to aid in transporting with the presence of HAs was introduced by He et al. (2014). A HA carrying a GPS and radio transmitter will have his pose tracked

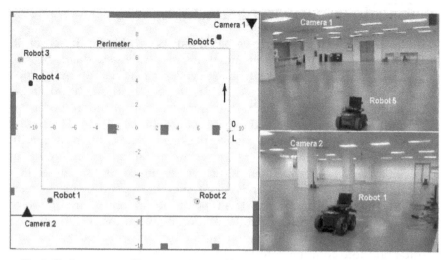

Fig. 9. Perimeter surveillance experimented by a team of robots (Acevedo et al., 2013).

by a RA. The radio transmitter will keep broadcasting HA's data. Another study regarding path partition strategy was investigated by Acevedo et al. (2013) that performed a perimeter surveillance with a team of mobile robots. Each robot should cover a specific non-overlapping section of the total path covered by the whole team. Each robot propagated its spatial data within its communication coverage range and a decentralized control algorithm coordinated the robots, based on the exchanged data between neighboring agents. The whole team collaborated to cover the whole area by monitoring each other's position and maintaining their subtasks as shown in Fig. 9.

3.2 Modest Cooperation Behaviour

A modest cooperation behaviour is another form of collaborative systems which requires establishing an explicit communication and implicit communication between agents. Both explicit communication and implicit are a via message data communication. However, the explicit communication message's data have a specific set of instructions such as requesting to dispatch a bale of hay, picking up fruit bins, or assisting another agent.

A cooperative agricultural system which consists of a supporting unit (SU) as a transporting vehicle to assist an operative agent as a primary unit (PU) was discussed by Jensen et al. (2012). They investigated the transporting unit navigation optimization and path planning which also involved in-field and inter-field transportation. The system was simulated

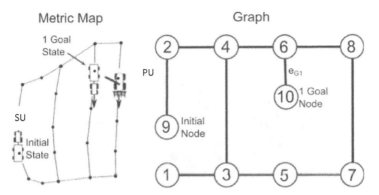

Fig. 10. The metric map and corresponding graph of a refilling of a PU unit simulation scenario (Jensen et al., 2012).

in MATLAB for a fertilizing operation when an autonomous tractor employed as a PU carrying a sprayer and required refilling its tank. The refilling operation was supported by a transporting vehicle working as a SU which arrives at the PU's location and refills as shown in Fig. 10.

3.3 Absolute Cooperation Behaviour

An absolute cooperation behaviour requires a continuous point to point implicit and explicit communication. An uninterrupted data exchange is established between collaborative agents such as pose correction of two agents carrying an object designed by Bailey et al. (2011). They designed a hybrid control to fuse inter-robot measurements. A distributed algorithm for joint localization of RAs enabled sharing spatial information between an advanced sensing RA with a lower sensing ability RA. The localization information sharing was obtained which continuously corrected agent's poses as a combined estimation method. The process allowed the advanced RA to help less equipped RAs. Each RA processed its own sensed data independently.

Another agricultural application which was discussed by Pitla et al. (2010) is grain harvesting as shown in Fig. 11. The application is a hybrid control system which had a central processing station (CMS) as a central controller and a grain harvesting robot GHR accompanied with two autonomous grain wagon robots GWR I and GWR II. GHR and GWR I are working together and keeping a close distance to each other in order to allow GHR to keep on feeding harvested hay to GWR I. The GWR II is to replace GWR I when it gets full.

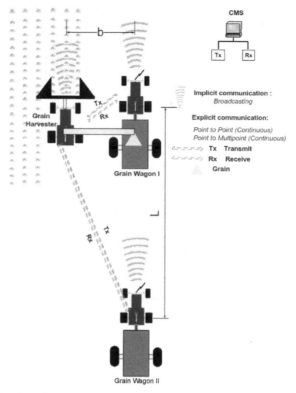

Fig. 11. Grain harvesting absolute cooperative system (Pitla et al., 2010).

4. MULTI-AGENT COOPERATIVE HYBRID CONTROL SYSTEM FRAMEWORK

In this section, the development of a Multi-agent Cooperative Hybrid Control System Framework for Agricultural Transportation Application in New Zealand is represented as a case study. This implementation investigates the optimization of the harvesting cost. A modest cooperative behaviour will be established within a team of RAs as fruit bins transporters, HAs as fruit pickers, and a computer base station as a central controller. The system structure is illustrated in Fig. 12. During a conventional harvesting process in commercial orchards, the picked fruits are collected in a fruit picking bag and emptied into a fixed bin placed on the orchard's row by the picker. Then, the bin when it gets full is transported by a forklift tractor to a collection station and inspected by a supervisor. The unproductive traveling time from and to a fixed bin prolongs the harvesting process.

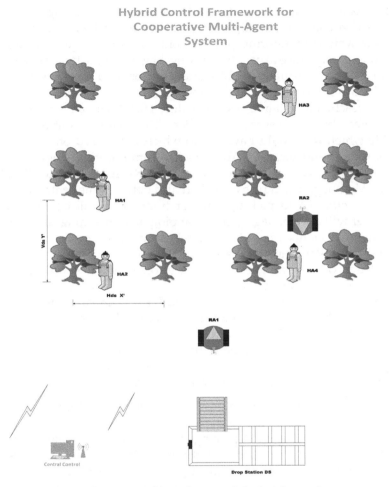

Fig. 12. Cooperative system framework for fruit harvesting.

An interactive monitoring of the human agents' movement and fruit picking process in an orchard is performed by the central Controller. Thus, an implicit communication message will exchange data between RAs and HAs which contains their identification number, location, time stamp, carried weight, and status. The central controller which is a computer is processing data broadcasted by each agent. It will process data and translate them into executable commands for RAs to act upon. The data processing will include noise filtering, segmentation, map matching, and weight monitoring which essentially contributes to the decision when a RA navigates to a HA's location and when the RA should navigate to a drop station. Therefore, a RA navigates to an HA's location when its picking bag is full or when requested while the RA will navigate to a drop station after collecting multiple full picking bags and its bin becomes full.

The weight monitoring process combines two stages, the first part is continuously measuring the picking bag's weight by load cells and broadcast the measured weight value, location, and timestamp as a data set via implicit communication messages, the second part is when the weight reached a certain threshold value. This threshold is contentiously monitored by the central controller. The central controller would decide which RA would navigate to which HA, based on their picking bag's weight and location. An explicit communication message will be established between a HA and the nearest RA requesting to dispatch a full picking bag. It will be established also between the central controller and a specific RA to pick up bags or transport a full bin. The central controller will also manage global mapping, path planning, and updating collaborative agents' maps. RAs are robust to the absence or failure of the central controller since they are decentralized controlled and can plan tasks based on the previously exchanged data and latest uploaded maps by the central controller.

A site visit was made to a Plant and Food Company's apple orchard in Motueka, New Zealand to investigate apples' harvesting process and to explore the possibility of automating bins' transportation task. It was concluded that automating the transportation process during harvesting is achievable because of the existing commercial and standard procedures followed by the apple plantation industry such as;

1. Fruit pickers move within groups in systematic patterns from end to end in each block visiting each tree once during harvesting. Apples are picked into carried picking bags which are emptied to a fixed bin located in the middle of the row. This procedure makes HAs movement patterns predictable by a central controller.

2. A commercial orchard consists of 7–12 blocks. Each block will have about 700–1100 apple trees depending on row plantation density and it has access from both ends allowing vineyard tractors to be driven through rows to perform spraying, fertilizing, transporting bins, etc. which will also allow RAs to access and exit easily and no need for complicated manoeuvres during navigation.

3. An apple tree with the height of 3 meters produces about 300–400 apples and a normal picker would pick 2–3 tons in a single day while a skilled picker can pick up to 4 tons. This massive production of apples encouraged automating the transportation process.

4. Movement between apples' trees are restricted since they are being supported by poles and steel wires; thus, pickers and tractors move through rows with a path clearance of 3 to 4 meters as shown in Fig. 13,

Fig. 13. Orchard's block and its two ends access.

therefore, RAs' navigation is going to be straightforward and easily managed.

5. Continuous monitoring of picking bags' weight will enable collecting data from every HA and improve harvesting productivity and quality.
6. Most orchards in New Zealand have internet service providers or cell phone coverage and have a good infrastructure.

Several algorithms would be integrated into the system in order to automate the process. Based on two level control hierarchy structure as shown in Fig. 14. A hybrid control has two collaborative algorithms layers which are, the first algorithms layer is cooperative algorithms processing received data from each agent, make and continuously update global maps, tasks' decision making, and navigation planning. The second algorithms layer is based on the RA processor as a decentralized controller that follows central controller's decision, take decisions during the absence of the central controller and execute local navigation. At the central controller layer, sorting algorithms will sort received data based on agents' IDs, carried weight, and position. Another search algorithm will locate the most suitable RA to assist an HA who has their picking bag full or nearly full. Locating the most suitable RA will be handled by a search algorithm based on two constraints which are the relative RA's location if available and the second is when all RAs are engaged, the search algorithm would look for a RA which will finish first to handle the request. The Dijkstra's algorithm is applied to plan the shortest path to be

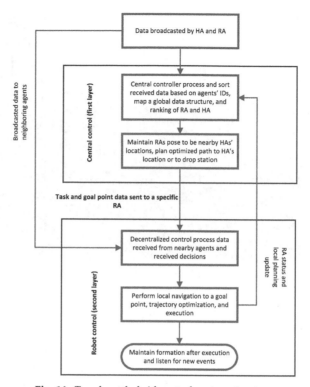

Fig. 14. Two-layer hybrid control system structure.

followed by the selected RA in order to reach the HA's or drop station's location.

On the robot control layer, RAs receive instructions from the central controller and data from neighboring agents and perform local navigation to a goal point including trajectory optimization and collision avoidance. RA would build its own global map and build a smaller data structure of other agents in order to enable robustness to central controller's failures and collaborate effectively with the rest of the team to execute commands.

5. FUTURE WORK

Multi-agent cooperative systems are current and evolving research areas and several aspects need to be investigated. These opportunities are the framework modules, the whole system's parts integration, operative algorithms, hardware's design, prototyping developments, experimental implementations, and evaluation studies.

One of the main challenges when applying a multi-agent systems is the prediction of HA's mobility patterns. It has a crucial role during harvesting and fruit transportation automation in this research. There are a few actively ongoing research studies in the arena of uncertainty and unpredicted behaviour of human's movement. Human mobility tracking in an agricultural process is complex and requires selective data mining operations. Therefore, future research should focus on improving algorithms and methodologies for monitoring unpredicted human mobility in agricultural applications.

Navigation and localization of mobile robots are also active research areas. There are numerous topics being studied to improve multi-robot navigation under real application constraints such as challenging terrains, weather condition, surface texture, and design complexity. In addition, navigation and localization algorithms development is another research field, even though, many algorithms have already been developed and optimized but an avenue for future studies to optimize new and fast algorithms that improve multi-agent cooperative navigation still exists.

The Multi-agent Cooperative Hybrid Control System Framework for Agricultural Transportation Application proposed in this chapter is an endeavor to improve the collaborative system's heterogeneity and optimization by including a HA as a functional team member. This design took an advantage of the existing standard mobility patterns in a commercial apple orchard to track HA's mobility. The proposed system discussed the implementation of the RAs as Unmanned Ground Vehicles (UGV) with a central controller. Future research can include quadrotors or rovers as Unmanned Aerial Vehicles (UAV) to improve communication coverage, global mapping, machine vision, and prediction human agent's mobility and which also can interact with one robot or several robots.

6. CONCLUSION

This chapter leads the reader through a review of cooperative multi-agent system applications in agriculture. The multi-agent collaborative concepts were introduced. The need for multi-agent systems in agriculture with existing designs and similar applications' backgrounds was illustrated. The review highlighted the effectiveness of automating certain agricultural processes and also discussed design considerations such as control topology, middleware and tools, navigation, and cooperative behaviour. Previously reviewed collaborative systems support these consideration and provide practical assessments for future designs.

HA to RA collaboration is a critical aspect of agriculture due to the tasks' complexity which requires humans' flexibility and intelligence. Thus, HA's role is not only to supervise robots exclusively but also to participate collaboratively and effectively communicate with the whole system. The unpredictable HA's mobility and tracking constraints is the reason for researchers' reluctance to employ HAs within cooperative multi-agent systems. The proposed study of the Reliable Multi-agent Cooperative Hybrid Control System Framework for Agricultural Transportation Application in New Zealand introduced HA's participation as an operative agent while improving a methodology for accurate mobility prediction by investigating the movement patterns during a harvesting process. A weight and pose tracking device is used to broadcast picking bags' weight and location which is received and processed by a central controller.

This chapter has shown possibilities and advantages of using cooperative multi-agent systems in agricultural applications. Selecting the applicable system design, level of cooperation, HA's and RA's task, and appropriate communication enhances RA's awareness and improves collaboration. Therefore, the development of multi-agent cooperation system in agricultural applications will be improved when RA and HA behave in more collaborative manners.

REFERENCES

Acevedo, J. J., Arrue, B. C., Maza, I., and Ollero, A. 2013. Cooperative perimeter surveillance with a team of mobile robots under communication constraints. Paper presented at the Intelligent Robots and Systems (IROS), 2013 IEEE/RSJ International Conference on.

Adamides, G., Berenstein, R., Ben-Halevi, I., Hadzilacos, T., and Edan, Y. 2012. User interface design principles for robotics in agriculture: The case of telerobotic navigation and target selection for spraying. Proceedings of the 8th Asian Conference for Information Technology in Agriculture, Taipei, Taiwan, 36.

Bailey, T., Bryson, M., Mu, H., Vial, J., McCalman, L., and Durrant-Whyte, H. 2011. Decentralised cooperative localisation for heterogeneous teams of mobile robots. Paper presented at the Robotics and Automation (ICRA), 2011 IEEE International Conference on.

Barca, J. C., and Sekercioglu, Y. A. 2013. Swarm robotics reviewed. Robotica. 31(03): pp. 345–359. doi:doi:10.1017/S026357471200032X.

Barth, R., Baur, J., Buschmann, T., Edan, Y., Hellström, T., Nguyen, T., Ringdahl, Ola, Saeys, Wouter, Salinas, Carlota, and Vitzrabin, R. 2014. Using ROS for agricultural robotics—design considerations and experiences. Paper presented at the Proceedings of the Second International Conference on Robotics and Associated High-Technologies and Equipment for Agriculture and Forestry, Madrid, Spain.

Bayındır, L. 2016. A review of swarm robotics tasks. Neurocomputing. 172: pp. 292–321. doi:http://dx.doi.org/10.1016/j.neucom.2015.05.116.

Cao, Y. U., Fukunaga, A. S., and Kahng, A. 1997. Cooperative Mobile Robotics: Antecedents and Directions. Autonomous Robots. 4(1): pp. 7–27. doi:10.1023/a:1008855018923.

Capodieci, N., and Cabri, G. 2013, 20–24 May. Collaboration in swarm robotics: A visual communication approach. Paper presented at the 2013 International Conference on Collaboration Technologies and Systems (CTS).

Cheng, L., Hou, Z.-G., and Tan, M. 2008. Decentralized adaptive leader-follower control of multi-manipulator system with uncertain dynamics. Paper presented at the Industrial Electronics, 2008. IECON 2008. 34th Annual Conference of IEEE.

Dias, M. B., and Stentz, A. 2000. A market approach to multirobot coordination (No. CMU-RI-TR-01-26). Carnegie-Mellon Univ Pittsburgh Pa Robotics Inst.

Emmi, L., Gonzalez-de-Soto, M., Pajares, G., and Gonzalez-de-Santos, P. 2014. New trends in robotics for agriculture: integration and assessment of a real fleet of robots. The Scientific World Journal, 2014.

English, A., Ball, D., Ross, P., Upcroft, B., Wyeth, G., and Corke, P. 2013. Low cost localisation for agricultural robotics. Proceedings of the 2013 Australasian Conference on Robotics & Automation.

Farangis Khosro, A., Richinder Singh, R., Fadi, A. F., Kent, D. W., and Stavros George, V. 2014. Sensor-based Stooped Work Monitoring in Robot-aided Strawberry Harvesting. Paper presented at the 2014 Montreal, Quebec Canada July 13–July 16, 2014, St. Joseph, Mich. http://elibrary.asabe.org/abstract.asp?aid=44962&t=5.

Fidan, B., Yu, C., and Anderson, B. 2007. Acquiring and maintaining persistence of autonomous multi-vehicle formations. IET Control Theory & Applications. 1(2): pp. 452–460.

Garro, B. A., Sossa, H., and Vazquez, R. A. 2007. Evolving ant colony system for optimizing path planning in mobile robots. Paper presented at the Electronics, Robotics and Automotive Mechanics Conference (CERMA 2007).

Gautam, A., and Mohan, S. 2012, 6–9 Aug. A review of research in multi-robot systems. Paper presented at the Industrial and Information Systems (ICIIS), 2012 7th IEEE International Conference on.

He, L., Arikapudi, R., Anjom, F. K., and Vougioukas, S. G. 2014. Worker position tracking for safe navigation of autonomous orchard vehicles using active ranging. Paper presented at the American Society of Agricultural and Biological Engineers Annual International Meeting 2014, ASABE 2014, July 13, 2014–July 16, 2014, Montreal, QC, Canada.

Holland, O., and Melhuish, C. 1999. Stigmergy, Self-Organization, and Sorting in Collective Robotics. Artificial Life. 5(2): pp. 173–202. doi:10.1162/106454699568737.

Jensen, M. A. F., Bochtis, D., Sørensen, C. G., Blas, M. R., and Lykkegaard, K. L. 2012. In-field and inter-field path planning for agricultural transport units. Computers and Industrial Engineering. 63(4): pp. 1054–1061. doi:http://dx.doi.org/10.1016/j.cie.2012.07.004.

Kazmi, W., Bisgaard, M., Garcia-Ruiz, F., Hansen, K. D., and la Cour-Harbo, A. 2011. Adaptive surveying and early treatment of crops with a team of autonomous vehicles. Paper presented at the European Conference on Mobile Robots.

Keicher, R., and Seufert, H. 2000. Automatic guidance for agricultural vehicles in Europe. Computers and Electronics in Agriculture. 25(1): 169–194.

Khan, S., and Ahmmed, M. K. 2016, 13–14 May. Where am I? Autonomous navigation system of a mobile robot in an unknown environment. Paper presented at the 2016 5th International Conference on Informatics, Electronics and Vision (ICIEV).

Li, N., Remeikas, C., Xu, Y., Jayasuriya, S., and Ehsani, R. 2015. Task assignment and trajectory planning algorithm for a class of cooperative agricultural robots. Journal of Dynamic Systems, Measurement and Control, Transactions of the ASME. 137(5). doi:10.1115/1.4028849.

Linz, A., Ruckelshausen, A., Wunder, E., and Hertzberg, J. 2014. Autonomous service robots for orchards and vineyards: 3D simulation environment of multi sensor-based navigation and applications. Paper presented at the 12th International Conference on Precision Agriculture, ISPA International Society of Precision Agriculture, Ed., Sacramento, CA, USA.

Ma, M., and Yang, Y. 2005. Adaptive triangular deployment algorithm for unattended mobile sensor networks. Paper presented at the International Conference on Distributed Computing in Sensor Systems.

Min Yang, J., Deguet, A., and Kazanzides, P. 2010, 18–22 Oct. A component-based architecture for flexible integration of robotic systems. Paper presented at the 2010 IEEE/RSJ International Conference on Intelligent Robots and Systems.

Mohamed, N., Al-Jaroodi, J., and Jawhar, I. 2008, 21–24 Sept. 2008. Middleware for Robotics: A Survey. Paper presented at the 2008 IEEE Conference on Robotics, Automation and Mechatronics.

Mousazadeh, H. 2013. A technical review on navigation systems of agricultural autonomous off-road vehicles. Journal of Terramechanics. 50(3): pp. 211–232. doi:http://dx.doi.org/10.1016/j.jterra.2013.03.004.

Noguchi, N., Will, J., Reid, J., and Zhang, Q. 2004. Development of a master–slave robot system for farm operations. Computers and Electronics in Agriculture. 44(1): pp. 1–19. doi:http://dx.doi.org/10.1016/j.compag.2004.01.006.

Pentjuss, A., Zacepins, A., and Gailums, A. 2011. Improving precision agriculture methods with multiagent systems in Latvian agricultural field. Engineering for rural development, Latvia, Jelgava, 109–114.

Pitla, S. K., Luck, J. D., and Shearer, S. A. 2010. Multi-Robot System Control Architecture (MRSCA) for Agricultural Production. 2010 Pittsburgh, Pennsylvania, June 20–June 23, 2010, 1.

Rodrigues, N., Pereira, A., and Leitão, P. 2013. Adaptive multi-agent system for a washing machine production line. pp. 212–223. *In*: Mařík, V., Lastra, J. L. M., and Skobelev, P. (eds.). Industrial Applications of Holonic and Multi-Agent Systems: 6th International Conference, HoloMAS 2013, Prague, Czech Republic, August 26–28, 2013. Proceedings. Berlin, Heidelberg: Springer Berlin Heidelberg.

Sharifi, M., Chen, X., and Pretty, C. G. 2016, 29–31 Aug. Experimental study on using visual odometry for navigation in outdoor GPS-denied environments. Paper presented at the 2016 12th IEEE/ASME International Conference on Mechatronic and Embedded Systems and Applications (MESA).

Sharifi, M., Young, M. S., Chen, X., Clucas, D., Pretty, C., and Meng, W. 2016. Mechatronic design and development of a non-holonomic omnidirectional mobile robot for automation of primary production. Cogent Engineering. 3(1): pp. 1250431. doi:10.1080/23311916.2016.1250431.

Vougioukas, S., Fountas, S., Blackmore, S., and Tang, L. 2005. Combining reactive and deterministic behaviours for mobile agricultural robots. Operational Research. 5(1): pp. 153–163.

Vougioukas, S., Spanomitros, Y., and Slaughter, D. 2012. Dispatching and routing of robotic crop-transport aids for fruit pickers using mixed integer programming. Paper presented at the American Society of Agricultural and Biological Engineers Annual International Meeting 2012, July 29, 2012–August 1, 2012, Dallas, TX, United states.

Wang, Z., Gong, L., Chen, Q., Li, Y., Liu, C., and Huang, Y. 2016. Rapid Developing the Simulation and Control Systems for a Multifunctional Autonomous Agricultural Robot with ROS. Paper presented at the International Conference on Intelligent Robotics and Applications.

Wurr, A., and Anderson, J. 2004. Multi-agent trail making for stigmergic navigation. Paper presented at the Conference of the Canadian Society for Computational Studies of Intelligence.

Zheng, Y. 2015. Trajectory data mining: An overview. ACM Trans. Intell. Syst. Technol. 6(3): pp. 1–41. doi:10.1145/2743025.

7

Adaptive Min-max Model Predictive Control for Field Vehicle Guidance in the Presence of Wheel Slip

Xu Wang,* Javad Taghia and *Jay Katupitiya*

1. INTRODUCTION

Developing highly accurate automatic guidance of agricultural vehicles can bring in immense benefits to the agricultural industry by way of cost savings. For example, autonomous agricultural vehicles that can accurately follow predefined paths can be used to plant the crop and then repeatedly revisit the growing crop accurately for crop management. Crop management includes growth monitoring and fertilizer, herbicides and pesticide application. Highly accurate autonomous machines can apply fertilizers, herbicides, and pesticides with greater spatial precision leading up to plant level care instead of field level care bringing in significant cost savings due to reduced fertilizer and chemical usage. In addition, use of autonomous systems address the skilled operator shortage, reduce the labor costs and improve occupational health and safety standards of operators (Van Henten et al., 2003).

Ensuring accurate operation of autonomous agricultural vehicles for path tracking is a challenging and complex task. The primary reason is that these vehicles operate on rough terrain, which at times can be sloping

School of Mechanical and Manufacturing Engineering University of New South Wales, NSW Australia.
Email: j.taghia@unsw.edu.au; j.katupitiya@unsw.edu.au
* Corresponding author: xu.wang3@student.unsw.edu.au

and undulating (Janulevičius and Giedra, 2009). Moreover, they carry out ground engaging operations such as plowing. These conditions often lead to slips at the front and rear wheels in both lateral and longitudinal directions (Eaton et al., 2009). Wheel slip is the interaction between soil and wheels, and affected by tyres as well as speed of vehicles, terrain properties, and path curvatures. In agricultural applications, the accuracy of lateral offsets is required to be within the range of five centimetres with respect to the reference path, even if farm vehicles are moving on slope and undulating ground (Lenain et al., 2006). The slips are inevitable disturbances affecting the operation of autonomous vehicles and need to be considered in designing path tracking controllers. For example, as the experimental results shown in (Lenain et al., 2004), classical control without sliding, deviated the farm tractor from the desired path with the highest error of 30 cm during the slope. While errors during the curve were in the range of the lowest lateral offset 40 cm to the highest offset 60 cm, during path tracking in the curve. The effect brought by wheel slip in field is significant, and can not be ignored. There are a few other researches while also highlighted the significant issue of wheel slip (Raheman and Jha, 2007; Pranav et al., 2010; Bevly et al., 2002). In this paper, one of the key aim is to investigate wheel slip. The autonomous technologies presented in this paper can be employed in other industries as well such as road construction in civil engineering and mining and defense.

The first step towards controller development is modeling. Two types of models can be used for controller development, the kinematic models or the dynamic models. Though dynamic models are more complex, they are more accurate than kinematic models especially when the vehicles operate with high accelerations. Dynamic models are also more specific to a given system than kinematic models, which are more general and easy to use. However, it has been shown that, for vehicles operating at low speeds with low accelerations such as farm vehicles, the kinematic models are accurate enough for designing path tracking controllers (Werner et al., 2012). A number of kinematic models based on non-slip assumption have been derived. However, this is not a valid assumption in the agricultural environment because slip is significant and inevitable (Micaelli and Samson, 1993; Samson, 1993). Lenain et al. (2006) extended a kinematic model by incorporating a rear side slip angle, and a front side slip angle to take the slip effects into account. A more comprehensive kinematic model was introduced by Fang et al. (2004). In this model, a lateral slip velocity perpendicular to the velocity of the field vehicle is added to the front wheels as a steering bias. As a further extension, a kinematic model (Huynh et al., 2010) was derived with three slips—lateral slip velocity, longitudinal slip velocity and steering slip angle. Their kinematic model can be used to obtain an offset model with respect to a reference path, the offsets being the lateral distance offsets and the heading offsets.

In this paper, the offset model of Huynh et al. (2010) is used to design the path tracking controller. Note that the kinematic models are nonlinear and, therefore, nonlinear controllers are recommended. Nonlinear approaches such as sliding mode control (SMC) and back stepping control (BSC) have been used for controlling mobile vehicles in many industrial applications (Yu and Kaynak, 2009; Taghia et al., 2015). Both control methods are based on Lyapunov stability analysis, and they are robust control methods that perform successfully in the presence of uncertainties and disturbances (Krstic et al., 1995). Both BSC (Huynh et al., 2012; Fang et al., 2006) and SMC (Taghia and Katupitiya, 2013) are found to be sensitive to unmatched uncertainties in the system model.

A very promising control method for achieving high precision path tracking is Model Predictive Control (MPC) due to its receding optimization and predictive ability. MPC has been successfully used in many industrial applications such as oil-refining and power systems (Qin and Badgwell, 2003; Richalet, 1993; Arnold and Andersson, 2011). In the recent past, researchers have shown an interest in applying MPC to path tracking. While there is an abundance of satisfactory research results, the majority of them use the assumption of pure rolling without sliding (Backman et al., 2009; Yaonan et al., 2010). As emphasized before, this assumption is invalid when it comes to the control of field vehicles in farming environments. Moreover, classical MPC is not inherently robust (Garcia et al., 1989), therefore it is necessary to design controllers taking the wheel slips into account. The work presented by Backman et al. (2010), took into account the wheel slip and used extended Kalman filter to compensate for the slippage, however, this approach is not robust due to the assumption of the Gaussian distribution of slip, which is not a reliable assumption. Lenain et al. (2005, 2006) used an extended kinematic model with two slip angles representing front and rear slip to design a control law and then created a sliding estimation algorithm to obtain the two slip angles. The results show acceptable performance, however, the noise levels on the two estimated slip angles were problematic.

This paper proposes adaptive and robust control approaches in such a way that the slip measurement or estimation is not required. Proposed min-max MPC is a robust MPC method that considers all possible disturbances including the worst case (Campo and Morari, 1987). However, at times, this method may cause overcompensation because the worst case does not occur always. To avoid overcompensation, Scokaert et al. (1998) proposed a min-max feedback MPC control method for a linear system. Although this method leads to a better performance than min-max MPC, it is computationally more intense.

In this paper, we propose a new MPC method called adaptive min-max MPC (AMM-MPC) which can deal with inherent slip through adaptation. The proposed MPC is called adaptive min-max MPC (AMM-MPC). This controller can deal with inherent slip through adaptation. In AMM-MPC, first a cost function is defined and then, the min-max technique is applied based on the bounds of the disturbances. The control law is adapted based on the curvature of the reference path and path offset feedback as the adaptive part. First, the derivation and stability of AMM-MPC are presented. Then, the performance of the proposed MPC is compared with the classical MPC in kinematic simulations and dynamic simulations. The dynamic simulation platform is developed to create a more realistic environment incorporating slip phenomena. Then, AMM-MPC and classical MPC are implemented on a tractor in the field, and results are compared and discussed. To compare the performance of the proposed controller with other types of controllers, the experiments were extended to include comparison of AMM-MPC with a SMC, which is presented under the title "Robust Adaptive Controller Design" (Fang, 2004), and a BSC (Huynh et al., 2012).

The breakdown of sections in this paper are as follows. In Section 2, kinematic modeling and system description are presented. In Section 3, AMM-MPC's derivation and stability analysis are provided. Evaluation of AMM-MPC in kinematic simulations, in dynamic simulations and field experiments are presented and discussed in Section 4. Then, the proposed AMM-MPC is compared with SMC and BSC in field experiments and the results are discussed in Section 5. Finally, the paper is concluded in Section 6.

2. SYSTEM DESCRIPTION AND MODELING

In this section, the kinematic model of a farm vehicle incorporating lateral and longitudinal wheel slips is described and an offset model is derived.

2.1 System Description

The field vehicle model is simplified into a bicycle model where two steered wheels are represented by a single steered wheel along the longitudinal axis of the field vehicle. The field vehicle is driven by rear wheels with a longitudinal speed v and steered at the front wheel with a steering angle δ. The vehicle kinematic model is shown in Fig. 1 and related variables and parameters are shown in Table 1.

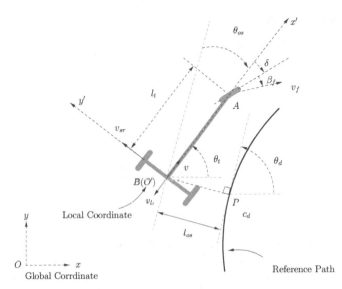

Fig. 1. Vehicle kinematic model and the reference path.

Table 1. Description of variables and points.

Variables	Description
c_d	curvature of the reference path at P
x_t	x coordinate of point O' in the global coordinate frame
y_t	y coordinate of point O' in the global coordinate frame
θ_t	orientation of the vehicle in the global coordinate frame
v	driving velocity vector at point B in the global coordinate frame, $v = \|\mathbf{v}\|$
$\mathbf{v}_f$	front wheel velocity vector, $v_f = \|\mathbf{v}_f\|$
θ_d	desired heading as per the reference path orientation
δ	front wheel steering angle
l_{os}	path offset
θ_{os}	heading offset
$\mathbf{v}_{sr}$	lateral slip velocity vector at B, $v_{sr} = \|\mathbf{v}_{sr}\|$
$\mathbf{v}_{lr}$	longitudinal slip velocity vector at B, $v_{lr} = \|\mathbf{v}_{lr}\|$
β_f	front wheel slip angle
l_t	vehicle wheelbase

Points	Description
A	center of the front axle
B	center of the rear axle
O	origin of global coordinate frame
O'	origin of local coordinate frame (coincides with B)
P	point of intersection of normal from B to the reference path

2.2 Kinematic Model

The field vehicle's states are defined by a vector $\mathbf{p}_t = [x_t,\ y_t,\ \theta_t]^T$. The kinematic equations for the field vehicle in the presence of wheel slips are derived using the kinematic model (Huynh et al., 2010),

$$\dot{x}_t = (v - v_{lr}) \cos \theta_t - v_{sr} \sin \theta_t,$$

$$\dot{y}_t = (v - v_{lr}) \sin \theta_t + v_{sr} \cos \theta_t, \tag{1}$$

$$\dot{\theta}_t = \frac{v - v_{lr}}{l_t} \tan(\delta + \beta_f) + \frac{v_{sr}}{l_t}.$$

2.3 Offset Model

The offset model is derived from the kinematic model in (1). Offset model consists of two states based on the position of the field vehicle with respect to the reference path, namely, the path offset l_{os} and the heading offset θ_{os}. The path offset l_{os} is defined as the distance $O'P$ in Fig. 1 while θ_{os} is defined as the angle $\theta_{os} = \theta_d - \theta_t$. Both l_{os} and θ_{os} are measurable based on the location of the vehicle obtained by RTK-GPS. The offset model is,

$$\dot{l}_{os} = -\sigma\,|\,v - v_{lr}\,|\, \sin \theta_{os} - \sigma\zeta\,v_{sr} \cos \theta_{os},$$

$$\dot{\theta}_{os} = \frac{v - v_{lr}}{l_t} \tan(\delta + \beta_f) + \frac{v_{sr}}{l_t} - \tag{2}$$

$$\sigma\,|\,v - v_{lr}\,|\,\frac{c_d \cos \theta_{os}}{1 + c_d l_{os}} + \sigma\zeta\,v_{sr}\frac{c_d \sin \theta_{os}}{1 + c_d l_{os}},$$

where σ is a direction coefficient. If σ is +1, the vehicle tracks the reference path in a clockwise direction. If σ is –1, the vehicle tracks the reference path in a counterclockwise direction. Another coefficient added to the model is ζ which is +1 when the vehicle moves forward and –1 when the vehicle moves backward. In this paper, the field vehicle is assumed to move forward only, and therefore ζ is always +1.

3. CONTROL DESIGN

In this section, AAM-MPC is derived and the stability analysis is presented.

As mentioned before, classical MPC is a successful control method when the model is accurate, however, in farming environments wheel slips are significant resulting in unsatisfactory performance of the path tracking controller. To manage significant disturbances in the field, an adaptive and robust AMM-MPC is derived, which is inspired by min-max MPC (Löfberg, 2003).

3.1 Feedback Linearization

The control design is based on the offset model in (2), and the objective of the control is to make the field vehicle follow the reference path accurately. Nevertheless, the offset model is highly nonlinear and directly using it in control design would be tedious. To simplify its use in control design, feedback linearization is carried out to convert the highly nonlinear system to a linear system (Khalil, 2002).

Assumption 3.1 *We assume $v > 0$ and $v > |v_{lr}|$, so we have*

$$\sigma|v - v_{lr}| = -\sigma(v - v_{lr}). \tag{3}$$

Assumption 3.1 is valid because we expect the tractor moves forward despite slipping.

Assumption 3.2 *It is feasible to linearize $\tan(\delta + \beta_f)$ so that we have,*

$$\tan(\delta + \beta_f) \approx \tan \delta + \tan \beta_f. \tag{4}$$

Assumption 3.2 is valid because the slip angle β_f is generally small in practical situations, usually between $0°$ and $5°$ (Huynh et al., 2012; Fang et al., 2006). With Assumptions 3.1 and 3.2, we simplify (2), and we use d_1 and d_2 to represent the overall disturbances. Then the model can be rewritten as,

$$\dot{l}_{os} = -\sigma v \sin \theta_{os} + d_1,$$
$$\dot{\theta}_{os} = \frac{v}{l_t} \tan \delta - \sigma v \frac{c_d \cos \theta_{os}}{1 + c_d l_{os}} + d_2, \tag{5}$$

where

$$d_1 = \sigma v_{lr} \sin \theta_{os} - \sigma v_{sr} \cos \theta_{os},$$

$$d_2 = -\frac{v_{lr}}{l_t} \tan \delta + \frac{v - v_{lr}}{l_t} \tan \beta_f + \frac{v_{sr}}{l_t} + \sigma v_{lr} \frac{c_d \cos \theta_{os}}{1 + c_d l_{os}} + \sigma v_{sr} \frac{c_d \sin \theta_{os}}{1 + c_d l_{os}}. \tag{6}$$

For canceling the nonlinearity in (5), we define two new state variables z_1, z_2 and one new control input u_k, as;

$$z_1 = l_{os},$$
$$z_2 = -\sigma v \sin \theta_{os}, \tag{7}$$

$$u_k = -\sigma v \cos \theta_{os} \left(\frac{v}{l_t} \tan \delta - \sigma v \, \frac{c_d \cos \theta_{os}}{1 + c_d l_{os}} \right).$$

Then the offset model becomes:

$$\dot{z}_1 = z_2 + \omega_1,$$
$$\dot{z}_2 = u_k + \omega_2, \tag{8}$$

where

$$\omega_1 = d_1,$$
$$\omega_2 = -\sigma v \cos \theta_{os} d_2. \tag{9}$$

Now, we define two vectors $z_k = [z_1 \; z_2]^T$ and $\omega_k = [\omega_1 \; \omega_2]^T$ so that a linear model is obtained as,

$$\dot{z}_k = A_c z_k + B_c u_k + D_c \omega_k,$$
$$y_k = C_c z_k. \tag{10}$$

where

$$A_c = \begin{bmatrix} 0 & 1 \\ 0 & 0 \end{bmatrix},$$

$$B_c = \begin{bmatrix} 0 \\ 1 \end{bmatrix},$$

$$C_c = \begin{bmatrix} \alpha \, \mathrm{sign}(l_{os}) & \gamma \, \mathrm{sign}(\theta_{os}) \end{bmatrix},$$

$$D_c = \begin{bmatrix} 1 \\ 1 \end{bmatrix}.$$

We use y_k in (10) to represent path tracking errors as the outputs, so the objective is to make y_k as close as possible to zero, where y_k depends on the values of α and γ which are gains on l_{os} and θ_{os}, respectively. For instance, if $\alpha = 1$ and $\gamma = 0$, y_k contains only l_{os} contribution, so the controller solely sends l_{os} to zero. Note that, sign() in C_c gurantees the combination between l_{os} and θ_{os} not to be diminished when their signs are opposite.

3.2 Augmented Model

The use of augmented models is to model uncertainties as disturbances acting on the system. The model in (10) is a continuous-time model and needs to be discretized as,

$$z_{k+1} = A_d z_k + B_d u_k + D_d \omega_k,$$

$$y_k = C_d z_k, \tag{11}$$

where A_d, B_d, C_d and D_d are discrete values for A_c, B_c, C_c and D_c.

We convert the linear state-space model in (11) to an augmented model with an embedded integrator (Wang, 2009). We define

$$\Delta z_k = z_k - z_{k-1},$$

$$\Delta u_k = u_k - u_{k-1}, \tag{12}$$

$$\Delta \omega_k = \omega_k - \omega_{k-1}$$

and we obtain the augmented model as,

$$\begin{bmatrix} \Delta z_{k+1} \\ y_{k+1} \end{bmatrix} = \begin{bmatrix} A_d & o_d^T \\ C_d A_d & 1 \end{bmatrix} \begin{bmatrix} \Delta z_k \\ y_k \end{bmatrix} + \begin{bmatrix} B_d \\ C_d B_d \end{bmatrix} \Delta u_k + \begin{bmatrix} D_d \\ C_d D_d \end{bmatrix} \Delta \omega_k,$$

$$y_k = \begin{bmatrix} o_d & 1 \end{bmatrix} \begin{bmatrix} \Delta z_k \\ y_k \end{bmatrix} \tag{13}$$

where $o_d = [0\ 0]$. To simplify, we define $x_k = [\Delta z_k^T\ y_k]^T$ and rewrite (13) as,

$$x_{k+1} = A x_k + B \Delta u_k + D \Delta \omega_k,$$

$$y_k = C x_k, \tag{14}$$

where,

$$A = \begin{bmatrix} A_d & o_d^T \\ C_d A_d & 1 \end{bmatrix},$$

$$B = \begin{bmatrix} B_d \\ C_d B_d \end{bmatrix},$$

$$C = \begin{bmatrix} o_d & 1 \end{bmatrix},$$

$$D = \begin{bmatrix} D_d \\ C_d D_d \end{bmatrix}.$$

In (14), $x_k \in \mathbb{R}^{3 \times 1}$, $y_k \in \mathbb{R}^{1 \times 1}$, $\Delta u_k \in \mathbb{R}^{1 \times 1}$, $\Delta \omega_k \in \mathbb{R}^{2 \times 1}$ denote the state, the controlled output, the augmented control input and the external disturbances, respectively.

The stability of the augmented system can be seen by the eigen values of the characteristic equation of matrix A as,

$$\det(\lambda I - A) = \det \begin{bmatrix} \lambda I - A_d & o_d^T \\ -C_d A_d & \lambda - 1 \end{bmatrix} \tag{15}$$

$$= (\lambda - 1) \det(\lambda I - A_d),$$

where I is an identity matrix. We can see that one obvious eigen value of A is 1 and other eigen values are decided by eigen values of matrix A_d. Note that the eigen value of 1 is the result of the integrator introduced to form the augmented model.

3.3 Adaptive Min-max Model Predictive Control Law

The novel contribution of this paper is the adaptive min-max MPC. The goal of this section is to derive the adaptive min-max MPC controller that explicitly considers the external disturbances based on (Wang et al., 2016). The "max" in min-max refers to worst-case scenarios. However, it can cause overcompensation. The adaptation introduced will avoid the overcompensation and will make it perform as close as possible to the classical nominal MPC formulation and at the same time tackle the disturbances.

The basic idea of predictive control is to calculate the future outputs together with the future control inputs by using the current states that are measurable. Objective function is minimized to obtain the optimal control trajectory, however, as, per MPC method, only the first control input is applied to the physical system. To begin with, we assume that at the sampling time k, $k \geq 0$, the current state is $x_{k|k}$, which is the same as x_k. Using $x_{k|k}$, the future states are predicted for N_p sample times which the prediction horizon. The state $x_{k+n|k}$ denotes the predicted state at $k + n$, predicted using $x_{k|k}$ at sampling instant k. The number of control inputs to obtain the future outputs are N_c which is the control horizon. Note that, $N_p \geq N_c$, preferably, $N_p > N_c$.

To obtain a simple notation, we introduce vectors to denote future states X, future outputs Y, future control inputs ΔU and unknown disturbances ΔW as,

$$X = (x_{k+1|k} \, x_{k+2|k} \, x_{k+3|k} \cdots x_{k+Np|k})^T$$

$$Y = (y_{k+1|k} \, y_{k+2|k} \, y_{k+3|k} \cdots y_{k+Np|k})^T$$

$$\Delta U = (\Delta u_{k|k} \, \Delta u_{k+1|k} \, \Delta u_{k+2|k} \cdots \Delta u_{k+Nc-1|k})^T$$

$$\Delta W = (\Delta \omega_{k|k} \, \Delta \omega_{k+1|k} \, \Delta \omega_{k+3|k} \cdots \Delta \omega_{k+Np-1|k})^T.$$

Then we can obtain,

$$Y = Fx_{k|k} + \Phi \Delta U + \Lambda \Delta W, \tag{16}$$

where

$$F = \begin{bmatrix} CA \\ CA^2 \\ CA^3 \\ \vdots \\ CA^{Np} \end{bmatrix},$$

$$\Phi = \begin{bmatrix} CB & 0 & 0 & \cdots & 0 \\ CAB & CB & 0 & \cdots & 0 \\ CA^2B & CAB & CB & \cdots & 0 \\ \vdots & \vdots & \vdots & \ddots & \vdots \\ CA^{Np-1}B & CA^{Np-2}B & CA^{Np-3}B & \cdots & CA^{Np-Nc}B \end{bmatrix},$$

$$\Lambda = \begin{bmatrix} CD & 0 & 0 & \cdots & 0 \\ CAD & CD & 0 & \cdots & 0 \\ CA^2D & CAD & CD & \cdots & 0 \\ \vdots & \vdots & \vdots & \ddots & \vdots \\ CA^{Np-1}D & CA^{Np-2}D & CA^{Np-3}D & \cdots & CD \end{bmatrix}.$$

The objective of model predictive control is to find the optimal ΔU such that the predicted output Y is as close as possible to the reference R_s. This process is implemented by minimizing a cost function J defined as,

$$J = (R_s - Y)^T (R_s - Y) + \Delta U^T \bar{R} \Delta U. \tag{17}$$

Then achieve,

$$\min_{\Delta U} J \text{ subject to}$$

$$\Delta U \in \Delta U^*, \tag{18}$$

where ΔU^* is constraint set matrix, $Y \in \mathbb{R}^{Np \times 1}$ and $\Delta U \in \mathbb{R}^{Nc \times 1}$. For path tracking, R_s is always set to 0, as offsets are driven to zero. Moreover, the diagonal matrix $\bar{R}$ is defined as $\bar{R} = r_w I_{Nc \times Nc}$ where $r_w \geq 0$ is a tuning parameter for penalizing the control input.

Next step is the management of the disturbances in the cost function, which is handled by a min-max method (Löfberg, 2003). Disturbances in the model (2) are all physical variables, thus they can be considered to have bounded values. Therefore we define,

$$\sup ||v_{lr}|| \leq v_{lr}^*,$$

$$\sup ||v_{sr}|| \leq v_{sr}^*,$$

$$\sup ||\beta_f|| \leq \beta_f^*,$$

where v_{lr}^*, v_{sr}^* and β_f^* are the bounds of the uncertainties, which are in fact the bounds of slip values. Now, substituting these bounds in (6) and (9), we obtain,

$$\sup | \, | \omega_1 | \, | \leq \omega_1^*,$$

$$\sup | \, | \omega_2 | \, | \leq \omega_2^*,$$

$$\omega_k \in \omega_k^*,$$

$$\Delta \omega_k \in \Delta \omega_k^*,$$

$$\Delta W \in \Delta W^*,$$

where ω_k^* and $\Delta \omega_k^*$ are bounded vectors of disturbances, and ΔW^* is normally taken as a constant matrix that corresponds to the worst case scenario.

The idea of min-max is to take the worst-case scenario into consideration, which is implemented by computing the cost function J using the bounds of external disturbances and then minimizing the cost function to obtain the optimal control input. Hence, the cost function J can be represented as,

$$\min_{\Delta U} \max_{\Delta W} Y^T Y + \Delta U^T \bar{R} \Delta U \text{ subject to}$$

$$\Delta U \in \Delta U^*, \tag{19}$$

$$\Delta W \in \Delta W^*.$$

However, this will cause overcompensation in most cases as the worst case cannot occur all the time. Therefore, it is very important to consider the field conditions to determine ΔW^*. For example, v_{sr} is a significant factor that makes the field vehicle deviate from the reference path, and it is generally larger during travel through high curvature segments of the path. However, it is insignificant during travel along straight segments. Thus, we can relate ΔW^* to (i) the curvature of the reference path to provide robustness and (ii) the amount of errors in offset values to provide adaptation. Hence, we define,

$$\Delta W^* = \frac{N_p}{[1\,1\,\cdots\,1]^T} (k_p c_d + k_q + k_t l_{os}), \tag{20}$$

where k_p is a value based on the worst case scenario when the curvature is not zero. The parameter k_q is a small positive constant at the worst case scenario representing zero curvature. The worst case scenario is decided by the bounds of v_{lr}, v_{sr} and β_f. The parameter k_t brings adaptive behavior, which is based on the amount of the path offset. The path offset is selected to contribute in the adaptive part of the controller due to the importance of the path offset in comparison to the heading offset in path tracking control.

Through minimization of J, the control trajectory vector ΔU is obtained, however, only the first control increment $\Delta u_{k|k}$ is applied as per MPC method, while other control inputs are ignored. Therefore,

$$\Delta u_k = \frac{\overbrace{N_c}}{[10 \cdots 0]^T} \Delta U$$

$$= -K_1 x_k - K_2, \tag{21}$$

where $K_1 = \frac{\overbrace{N_c}}{[10 \cdots 0]^T} (\Phi^T \Phi + \bar{R})^{-1}(\Phi^T F)$ and $K_2 = \frac{\overbrace{N_c}}{[10 \cdots 0]^T} (\Phi^T \Phi + \bar{R})^{-1}(\Phi^T \Lambda \Delta W)$.

Finally, from (7) and (12), the steering angle δ, which is the actual control input is calculated as,

$$u_k = \Delta u_k + u_{k-1},$$

$$\delta = \arctan\{(\frac{l_t}{v - v_{lr}})(u_k + \sigma |v| \frac{c_d \cos \theta_{os}}{1 + c_d l_{os}})\}. \tag{22}$$

3.4 Stability Analysis

The stability is proven using Lyapunov criterion based on the approach in Wang et al. (2009, 2016).

Theorem 3.3 *Given that the cost function J is minimized subjected to $\Delta U \in \Delta U^*$ and the constraint on the final output $y_{k+Np} = 0$ resulting from the control inputs $\Delta u_k, ...\Delta u_{k+Np-1}$, the closed loop MPC system is asymptotically stable.*

Proof: From Subsection 3.3, we know that AMM-MPC is realized by receding optimization. The future control trajectory $\Delta u_k, ...\Delta u_{k+Np-1}$ at time k is optimized by minimizing the cost function J_k, represented as,

$$J_k = \sum_{i=1}^{Np} y_{k+i}^T y_{k+i} + \sum_{i=0}^{Np-1} \Delta u_{k+i}^T r_w \Delta u_{k+i}, \tag{23}$$

where J_k is subjected to constraints and $r_w \geq 0$ is a gain.

Now, we assume the Lyapunov function $V(x_k)$ is equal to the minimum of the cost function J_k with the optimal control trajectory $\Delta u_k, ..., \Delta u_{k+Np-1}$ and corresponding outputs $y_{k+1}, ..., y_{k+Np}$, represented as,

$$V(x_k) = \min J_k$$

$$= \sum_{i=1}^{Np} y_{k+i}^T y_{k+i} + \sum_{i=0}^{Np-1} \Delta u_{k+i}^T r_w \Delta u_{k+i}. \tag{24}$$

The Lyapunov function $V(x_k)$ at sampling instant k is positive definite and $V(x_k)$ is infinite if x_k is infinite. Similar to $V(x_k)$, the Lyapunov function $V(x_{k+1})$ at time $k+1$ is the minimum of the cost function J_{k+1} with the optimal control trajectory $\Delta u_{k+1}, ..., \Delta u_{k+Np}$ and corresponding outputs $y_{k+2}, ..., y_{k+Np+1}$, represented as,

$$V(x_{k+1}) = = \sum_{i=1}^{Np} y_{k+1+i}^T y_{k+1+i} + \sum_{i=0}^{Np-1} \Delta u_{k+1+i}^T r_w \Delta u_{k+1+i}. \tag{25}$$

Now a function $\bar{V}$ will be introduced to relate $V(x_k)$ to $V(x_{k+1})$. The optimal control trajectory of $V(x_k)$ is shifted one step forward and its last control input Δu_{k+Np} is set to zero. The function $\bar{V}$ is formed by evaluating $V(x_{k+1})$ at the above mentioned time shifted control trajectory, which is a non-optimal control trajectory. For any non-optimal control trajectory the objective function has to be greater or equal to $V(x_{k+1})$. Therefore,

$$V(x_{k+1}) \le \bar{V}. \tag{26}$$

Based on (14), $\bar{V}$ has the same control trajectory as $V(x_k)$ at sampling times $k+1, k+2, ..., k+N_p-1$, thus,

$$V(x_{k+1}) - V(x_k) \le \bar{V} - V(x_k), \tag{27}$$

then,

$$\bar{V} - V(x_k) = y_{k+Np}^T y_{k+Np} - y_{k+1}^T y_{k+1} - \Delta u_k^T r_w \Delta u_k. \tag{28}$$

Given that as per Theorem 3.3, $y_{k+Np} = 0$,

$$\bar{V} - V(x_k) = -y_{k+1}^T y_{k+1} - \Delta u_k^T r_w \Delta u_k. \tag{29}$$

Therefore, the derivative of the Lyapunov function is,

$$V(x_{k+1}) - V(x_k) \leqslant -y_{k+1}^T y_{k+1} - \Delta u_k^T r_w \Delta u_k < 0. \tag{30}$$

This proves the asymptotic stability of the closed-loop system.

4. EVALUATION AND COMPARISON OF CLASSICAL MPC AND AMM-MPC

4.1 Kinematic Simulation

In this section, the proposed AMM-MPC is used to control the kinematic model of the tractor to follow a predefined path and the performance is compared with the classical MPC (Wang, 2009).

A simulation platform has been developed to compare AMM-MPC with classical MPC. The kinematic model incorporates slip velocities in lateral and longitudinal directions. In the kinematic platform, the lateral slip velocity v_{sr} is considered to be 30 percent less than the field vehicle velocity and is generated by random numbers related to the curvature of the reference path as per $v_{sr} = -10\, c_d \cos \delta - 0.3\, v\, \delta$ (5 Rand() − 5), where Rand() is a uniform random number generator, generating numbers between 0 and 1. The longitudinal velocity v_{lr} is considered less than 30 percentage of the field vehicle velocity and calculated as per $v_{lr} = 0.3\, v$ (Rand() − 0.5) + sin δ. The steering slip angle β_f is defined as random number within the range −5° to 5°, and calculated using $\beta_f = 10$ (Rand() − 0.5).

For the controller, the matrix C_m is chosen as [1.5 0.75], that indicates the path offset is more important than the heading offset in the path tracking. Finally, we have a mechanical constraint for steering angle δ, which is defined as −45° ≤ δ ≤ 45°. Parameters for the tractor and the controller are listed in Table 2.

The reference path used in the simulation is shown in Fig. 2. The vehicle starts at the point marked by a star and runs in the clockwise direction. This path has straight and curved segments, and the curved segments have different curvatures.

The proposed controller, AMM-MPC, and the classical MPC are used to control the vehicle represented by a kinematic model. Path offsets and heading offsets are recorded for comparison.

In Fig. 3, the path offset obtained from AMM-MPC and classical MPC are plotted. As shown, at the corners the amount of path offset increased in classical MPC in contrast to AMM-MPC, which showed more consistent performance throughout the path. In Fig. 4, heading offsets for AMM-MPC and MPC are shown. As can be seen, the performance of AMM-MPC is better than classical MPC.

To provide a better quantitative comparison, box plots are shown for the absolute value of path offsets and heading offsets in Figs. 5 and 6, respectively. The red points indicate the outliers and the red lines in the middle are the medians, which is better when it is closer to zero. The upper and lower quartiles are shown as blue lines. The better performance of AMM-MPC can be seen in the box plots.

Table 2. Parameters for simulation and experiment.

Parameters	Value
l_t	1.7 m
v	3 m/s
r_w	0.1
N_p	5
N_c	2
α	1.5
γ	0.75

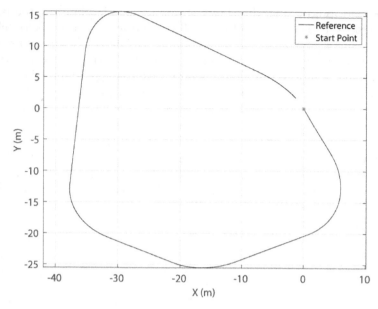

Fig. 2. The reference path used in the evaluation of AMM-MPC and comparison with classical MPC.

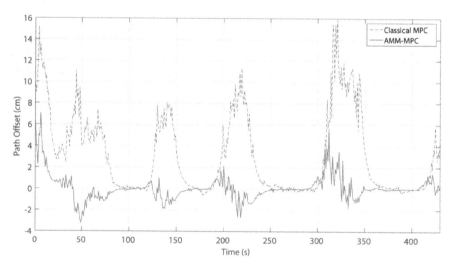

Fig. 3. Path offset for AMM-MPC and classical MPC in kinematic simulation.

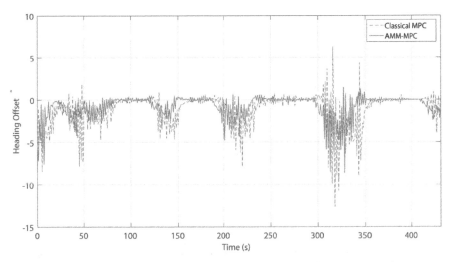

Fig. 4. Heading offset for AMM-MPC and classical MPC in kinematic simulation.

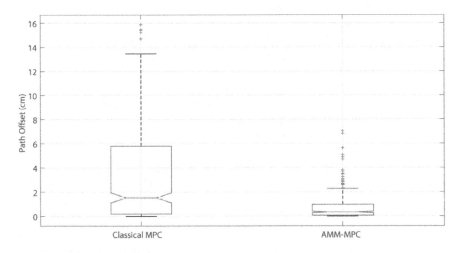

Fig. 5. Box plot for path offsets for AMM-MPC and classical MPC in kinematic simulation.

4.2 Dynamic Simulation

Slip is a result of forces due to wheel-ground interaction. To investigate the performance of the controllers in the presence of slip forces, a more realistic simulation environment was created including the dynamic model of a tractor incorporating a wheel model generating the slip forces.

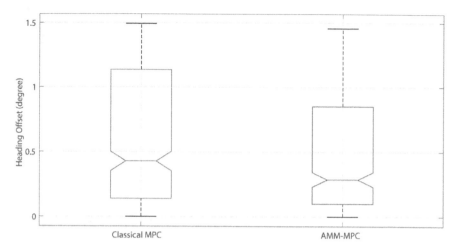

Fig. 6. Box plot heading offset for AMM-MPC and classical MPC in kinematic simulation.

Simulation is handled in C++ using the dynamic models of a tractor (Siew et al., 2009) and are solved using an explicit Runge-Kutta method (Trimbitas and Trimbitas, 2007) with a 8.33 μs time step. To model terrain uncertainty, a parametric noise map is introduced into the wheel-ground system in the form of simplex noise. Under each wheel, the contact surface is determined by evaluating the noise function across a small region of the contact patch, which is used to determine contact and slip forces in the wheel model based on the surface's up-vector direction. These noise characteristics are configurable in the simulation platform, in such a manner that slips and disturbances fall within a specified bounded range.

The reference path for the dynamic simulation is the same as the kinematic simulation, which is shown in Fig. 2. The path offset and the heading offset obtained from the dynamic simulation are plotted for AMM-MPC and classical MPC in Figs. 7 and 8, respectively. Similar to the kinematic simulation results, accuracy in the path following for AMM-MPC is significantly better at the corners in comparison to the classical MPC. Along the straight segments of the path, the improvements shown by the AMM-MPC are minor in comparison to the classical MPC.

Box plots are shown in Figs. 9 and 10 for the absolute value of path offsets and the heading offsets, respectively. The plots confirm the significant improvements brought about by the proposed AMM-MPC.

4.3 Field Experiment

The controllers were implemented on the autonomous tractor shown in Fig. 11. The tractor is a John Deere 4210 Compact Utility Tractor and

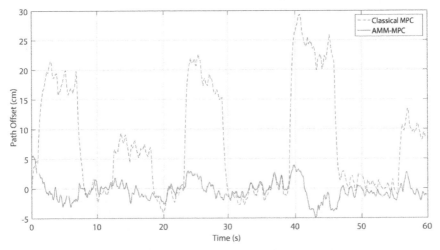

Fig. 7. Path offset for AMM-MPC and classical MPC in dynamic simulation.

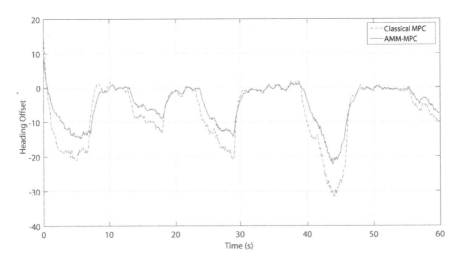

Fig. 8. Heading offset for AMM-MPC and classical MPC in dynamic simulation.

was made an autonomous vehicle at the University of New South Wales, Australia. More details about software and hardware can be found in Matveev et al. (2013); Eaton et al. (2008) and Taghia et al. (2015).

The same reference path shown in Fig. 2 is used for this field experiment. The path offset and heading offset values obtained from the field experiment are plotted in Figs. 12 and 13 respectively. As before, the performance of AMM-MPC is superior when compared with classical

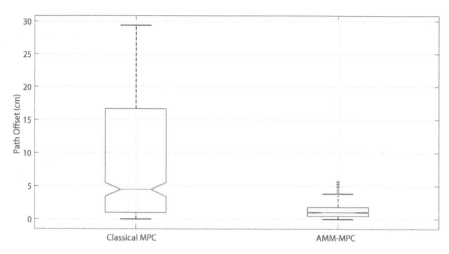

Fig. 9. Box plot path offset for AMM-MPC and classical MPC in dynamic simulation.

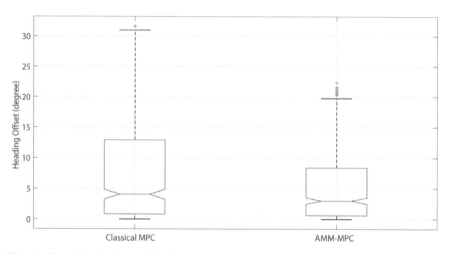

Fig. 10. Box plot heading offset for AMM-MPC and classical MPC based on dynamic simulation.

MPC, especially at the segments with higher curvature. This conclusion is supported by the box plots in Figs. 14 and 15.

As it is noticeable in Figs. 3, 7 and 12, AMM-MPC performed more accurately and robustly in both curved and straight segments of the path. In addition, AMM-MPC dealt with slip adaptively, without requiring slip estimation, which is the main advantage of the proposed method.

Fig. 11. John Deere 4210 Compact Utility Tractor used in field experiments.

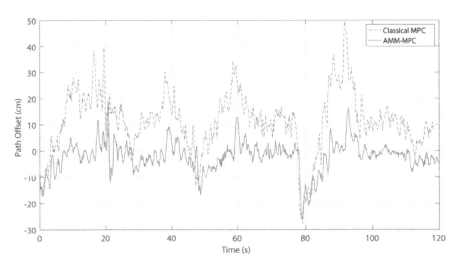

Fig. 12. Path offset for AMM-MPC and classical MPC in field experiments.

5. COMPARISON OF AMM-MPC WITH SMC AND BSC

Based on the results obtained from the kinematic simulation, the dynamic simulation and the field experiment in Section 4, it can be concluded that our approach in dealing with slip as uncertainty in AMM-MPC is successful. The robust and adaptive behavior of AMM-MPC has improved over that of

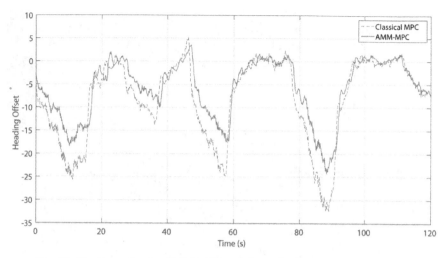

Fig. 13. Heading offset for AMM-MPC and classical MPC in field experiments.

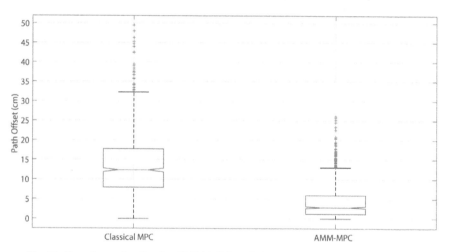

Fig. 14. Box plot path offset for AMM-MPC and classical MPC in field experiments.

the classical MPC significantly. Note that, there is no measurement of slip directly or indirectly, which makes the proposed controller a general and reliable path tracking method. To verify further, the proposed AMM-MPC is compared with a successful SMC implementation, which is presented under the title "Robust Adaptive Controller Design" (Fang, 2004), and a BSC reported in literature that showed good performance (Huynh et al., 2012).

In this field experiment, as the reference path, long farm path was selected. The path is shown in Fig. 16.

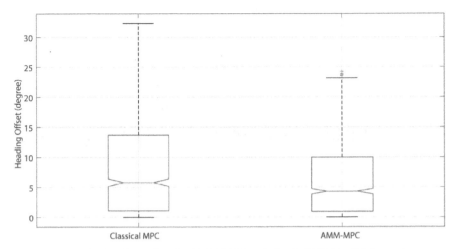

Fig. 15. Box plot heading offset for AMM-MPC and classical MPC in field experiments.

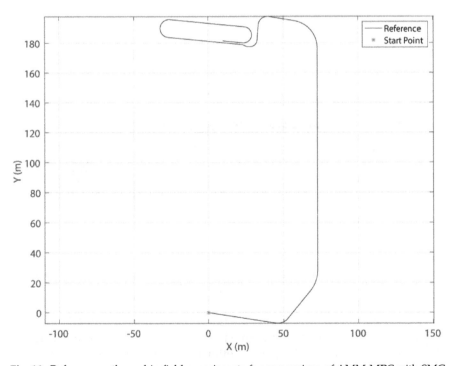

Fig. 16. Reference path used in field experiments for comparison of AMM-MPC with SMC and BSC.

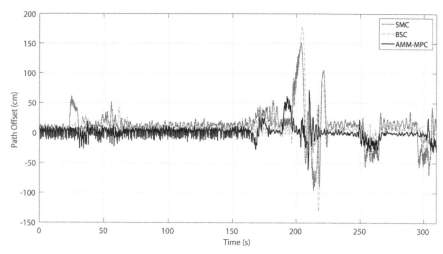

Fig. 17. Path offset in field experiments for comparison of AMM-MPC with SMC and BSC.

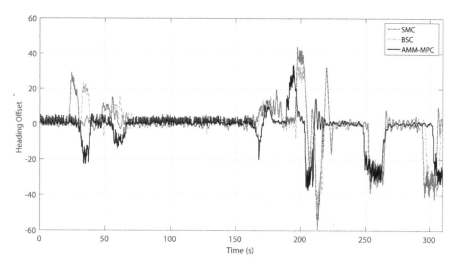

Fig. 18. Heading offset in field experiments for comparison of AMM-MPC with SMC and BSC.

The path offset values and heading offset values obtained from this experiment were recorded and are shown in Figs. 17 and 18 respectively. As we can see, the results show better performance with the proposed controller. Once again, improvement is more noticeable at the curved segments as shown, i.e., during 180 to 240 seconds.

For a more compact quantitative comparison, box plots and tables are presented showing absolute values, root mean square (RMS) values and

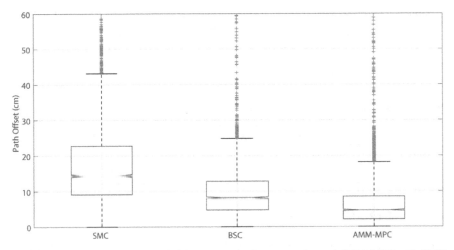

Fig. 19. Box plot of path offset in field experiments for comparison of AMM-MPC with SMC and BSC.

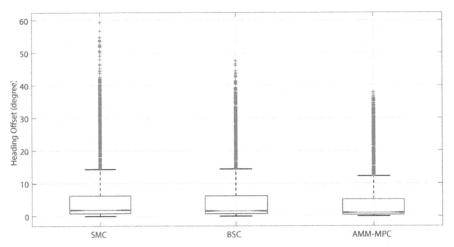

Fig. 20. Box plot of heading offset in field experiments for comparison of AMM-MPC with SMC and BSC.

standard deviation (SD) values for the three controllers. RMS value for path offset for AMM-MPC is about 12 cm that is significantly better than the respective values of BSC and SMC which are 26 cm and 30 cm. For the heading offset the difference is not significant. However, the heading accuracy is also better in AMM-MPC with RMS value of 10.44° and SD of 10.1° compared to those of SMC and BSC, shown in Tables 3 and 4.

Table 3. Path offset root mean square (RMS) values and standard deviation (SD) values.

Path Offset (cm)	SMC	BSC	AMM-MPC
RMS	30.33	26.88	11.42
SD	27.74	25.04	11.39

Table 4. Heading offset root mean square (RMS) values and standard deviation (SD) values.

Heading Offset (°)	SMC	BSC	AMM-MPC
RMS	12.48	11.50	10.44
SD	12.47	11.46	10.10

6. CONCLUSIONS

This paper proposed a very novel and promising adaptive min-max model predictive controller for path tracking control of farm vehicles in the presence of slip. The proposed controller's derivation and stability proof were presented. The performance of the proposed controller was evaluated with extensive simulation incorporating kinematic simulation, dynamic simulation and real field experiments in which the performance of the AMM-MPC was compared with classical MPC's performance. The proposed controller was also compared with two successful implementations of other forms of robust nonlinear controllers, namely, a sliding mode controller and a back stepping controller in field experiments on a typical farm. The results obtained show significant improvements in the accuracy in path offsets and heading offsets, especially at the segments with higher curvatures, where slip is greater. AMM-MPC not only provided robustness but also dealt with wheel slip adaptively without requiring slip measurement or estimation, which is the major contribution of the proposed controller.

7. ACKNOWLEDGMENTS

The authors wish to acknowledge Stanley Lam for making his Dynamic Systems Simulation Software available for the simulation study and Stephen Kuhle and Vincenco Carnivale for providing the necessary technical support for the research.

REFERENCES

Arnold, M., and Andersson, G. 2011. Model predictive control of energy storage including uncertain forecasts. In Power Systems Computation Conference (PSCC), Stockholm, Sweden.

Backman, J., Oksanen, T., and Visala, A. 2009. Parallel guidance system for tractor-trailer system with active joint. In 7th European Conference on Precision Agriculture, Wageningen, Netherlands, July 6–8, pp. 615–622.

Backman, J., Oksanen, T., and Visala, A. 2010. Nonlinear model predictive trajectory control in tractor-trailer system for parallel guidance in agricultural field operations. In International Conference on Agricontrol, Kyoto, Japan, December 6–8, 2.

Bevly, D. M., Gerdes, J. C., and Wilson, C. 2002. The use of GPS based velocity measurements for measurement of sideslip and wheel slip. Veh. Syst. Dyn. 38(2): pp. 127–147.

Campo, P. J. and Morari, M. 1987. Robust model predictive control. In American Control Conference, IEEE. pp. 1021–1026.

Eaton, R., Katupitiya, J., Pota, H., and Siew, K. W. 2009. Robust sliding mode control of an agricultural tractor under the influence of slip. In International Conference on Advanced Intelligent Mechatronics, IEEE/ASME. pp. 1873–1878.

Eaton, R., Katupitiya, J., Siew, K., and Dang, K. 2008. Precision guidance of agricultural tractors for autonomous farming. In 2nd Annual IEEE Systems Conference, Montreal, Canada. pp. 1–8.

Fang, H. 2004. Automatic guidance of farm vehicles in presence of sliding effects. A scientific report for the post-doc research project of all-terrain autonomous vehicle control, Le LASMEA.

Fang, H., Fan, R., Thuilot, B., and Martinet, P. 2006. Trajectory tracking control of farm vehicles in presence of sliding. Robot. Auton. Syst. 54(10): pp. 828–839.

Garcia, C. E., Prett, D. M., and Morari, M. 1989. Model predictive control: theory and practice-a survey. Autom. 25(3): pp. 335–348.

Huynh, V., Katupitiya, J., Kwok, N., and Eaton, R. 2010. Derivation of an error model for tractor-trailer path tracking. In International Conference on Intelligent Systems and Knowledge Engineering, IEEE. pp. 60–66.

Huynh, V., Smith, R., Kwok, N. M., and Katupitiya, J. 2012. A nonlinear PI and backstepping-based controller for tractor-steerable trailers influenced by slip. In International Conference on Robotics and Automation, IEEE. pp. 245–252.

Janulevičius, A., and Giedra, K. 2009. The slippage of the driving wheels of a tractor in a cultivated soil and stubble. Trasp. 24(1): pp. 14–20.

Khalil, H. K. 2002. Nonlinear Systems. New Jersey: Prentice Hall.

Krstic, M., Kokotovic, P. V., and Kanellakopoulos, I. 1995. Nonlinear and Adaptive Control Design, first ed. John Wiley & Sons, Inc., New York, NY, USA.

Lenain, R., Thuilot, B., Cariou, C., and Martinet, P. 2004. Non-linear control for car like mobile robots in presence of sliding: Application to guidance of farm vehicles using a single RTK GPS. In International Conference on Robotics and Automation (ICRA), New Orleans (USA). pp. 1873–1878.

Lenain, R., Thuilot, B., Cariou, C., and Martinet, P. 2005. Model predictive control for vehicle guidance in presence of sliding: application to farm vehicles path tracking. In International Conference on Robotics and Automation, IEEE. pp. 885–890.

Lenain, R., Thuilot, B., Cariou, C., and Martinet, P. 2006. High accuracy path tracking for vehicle in presence of sliding: application to farm vehicles automatic guidance for agricultural task. Auton. Robot. 21(1): pp. 79–97.

Löfberg, J. 2003. Minimax approaches to robust model predictive control, PhD thesis. Linköping University.

Matveev, A. S., Hoy, M., Katupitiya, J., and Savkin, A. V. 2013. Nonlinear sliding mode control of an unmanned agricultural tractor in the presence of sliding and control saturation. Robot. Auton. Syst. 61(9): pp. 973–987.

Micaelli, A. and Samson, C. 1993. Trajectory tracking for unicycle-type and two-steering-wheels mobile robots. INRIA Technical Report No. 2097.

Pranav, P., Pandey, K., and Tewari, V. 2010. Digital wheel slipmeter for agricultural 2WD tractors. Comput. Electron. Agric. 73(2): pp. 188–193.

Qin, S. J. and Badgwell, T. A. 2003. A survey of industrial model predictive control technology. Control Eng. Pract. 11(7): pp. 733–764.

Raheman, H. and Jha, S. 2007. Wheel slip measurement in 2WD tractor. J. Terramechanics. 44(1): pp. 89–94.

Richalet, J. 1993. Industrial applications of model based predictive control. Autom. 29(5): pp. 1251–1274.

Samson, C. 1993. Mobile robot control part 2: control of chained systems and application to path following and time-varying point-stabilization of wheeled vehicles. INRIA Technical Report No. 1994.

Scokaert, P., and Mayne, D. 1998. Min-max feedback model predictive control for constrained linear systems. IEEE Tran. Autom. Control. 43(8): pp. 1136–1142.

Siew, K., Katupitiya, J., Eaton, R., and Pota, H. 2009. Simulation of an articulated tractor-implement-trailer model under the influence of lateral disturbances. In International Conference on Advanced Intelligent Mechatronics, IEEE/ASME. pp. 951–956.

Taghia, J., and Katupitiya, J. 2013. Wheel slip identification and its use in the robust control of articulated off-road vehicles. In International Conference on Robotics and Automation, IEEE, Sydney, Australia.

Taghia, J., Wang, X., Lam, S., and Katupitiya, J. 2015. A sliding mode controller with a nonlinear disturbance observer for a farm vehicle operating in the presence of wheel slip. Auton. Robot. pp. 1–18.

Trimbitas, R. and Trimbitas, M. 2007. Runge-Kutta methods and inverse hermite interpolation. In International Symposium on Symbolic and Numeric Algorithms for Scientific Computing (SYNASC), IEEE. pp. 118–123.

Van Henten, E., Van Tuijl, B. v., Hemming, J., Kornet, J., Bontsema, J., and Van Os, E. 2003. Field test of an autonomous cucumber picking robot. Biosyst. Eng. 86(3): pp. 305–313.

Wang, L. 2009. Model predictive control system design and implementation using MATLAB®. London: Springer-Verlag.

Wang, X., Taghia, J., and Katupitiya, J. 2016. Robust model predictive control for path tracking of a tracked vehicle with a steerable trailer in the presence of slip. In 5th IFAC Conference on Sensing, Control and Automation Technologies for Agriculture, Seattle, WA, USA, pp. 469–474. Elsevier.

Werner, R., Muller, S., and Kormann, K. 2012. Path tracking control of tractors and steerable towed implements based on kinematic and dynamic modeling. In 11th International Conference on Precision Agriculture, Indianapolis, Indiana, USA. pp. 15–18.

Yaonan, W., Yimin, Y., Xiaofang, Y., Feng, Y., and Shuning, W. 2010. A model predictive control strategy for path-tracking of autonomous mobile robot using electromagnetism-like mechanism. In International Conference on Electrical and Control Engineering, IEEE. pp. 96–100.

Yu, X., and Kaynak, O. 2009. Sliding-mode control with soft computing: A survey. IEEE Trans. Ind. Electron. 56(9): pp. 3275–3285.

8

Model Reference Adaptive Control for Uncertain Dynamical Systems with Unmatched Disturbances

A Command Governor-Based Approach[#]

Ehsan Arabi, Tansel Yucelen* and Benjamin C. Gruenwald

1. INTRODUCTION

Numerous agriculture applications involving robotic and mechatronic systems require precise feedback control laws to accomplish given tasks with high accuracy. For example, tasks such as autonomous seeding, harvesting, and/or row cropping, unmanned ground vehicles have to precisely run parallel in the presence of disturbances and uncertainties resulting from variations in unknown ground frictions and potential unpredictable damages to the vehicle dynamics (Lenain et al., 2003; Cariou et al., 2009). Unmanned aerial vehicles, that have recently come in use for agriculture applications to maximize yields and minimize potential crop damages (Saari et al., 2011; Primicerio et al., 2012; Mäkynen et al., 2012; Tokekar et al., 2016), too need precision. It is often hard to autonomously operate these vehicles, especially the fixed-wing ones, in

Laboratory for Autonomy, Control, Information, and Systems Department of Mechanical Engineering, University of South Florida, USA.
* Corresponding author: Engineering Building C 2209, 4202 East Fowler Avenue, Tampa, Florida 33620, United States of America. Email: yucelen@lacis.team
\# This research was supported by the National Aeronautics and Space Administration under Grant NNX15AM51A.

challenging weather conditions in a precise close proximity to the ground (e.g., for farm imaging and monitoring) due to the increased uncertain lift forces and decreased uncertain aerodynamic drags (McRuer et al., 2014; Stengel, 2015). Thus, one of the fundamental problems arising in control technologies for autonomous agriculture vehicle applications is to achieve a level of desired, precise closed-loop system performance in the presence of a broad class of disturbances and uncertainties. To this end, model reference adaptive control architectures provide promising system stability and desired performance when the nature of the disturbances and uncertainties are matched in the system dynamics.

Yet, in many agriculture applications like the above ones when the matching assumption does not hold, the design of model reference adaptive control laws becomes a challenge. Notable model reference adaptive control contributions addressing this challenge include (Cao and Hovakimyan, 2008; Xargay et al., 2010; Li and Hovakimyan, 2012; Leman et al., 2010; Che and Cao, 2012; Stepanyan and Krishnakumar, 2015; Boskovic and Han, 2009; Stepanyan and Krishnakumar, 2012; Kristic et al., 1995; Heise and Holzapfel, 2015; Fravolini and Campa, 2011). In particular, the authors of (Cao and Hovakimyan, 2008; Xargay et al., 2010; Li and Hovakimyan, 2012; Leman et al., 2010; Che and Cao, 2012) use an adaptive control law based on a low-pass filter in the control channel and an estimation scheme. In the context of fault-tolerant adaptive flight control, the authors of (Stepanyan and Krishnakumar, 2015; Boskovic and Han, 2009; Stepanyan and Krishnakumar, 2012; Kristic et al., 1995) use certainty equivalence adaptive control as an indirect adaptive control design method for systems with unmatched uncertainties. The authors of (Heise and Holzapfel, 2015; Fravolini and Campa, 2011) proposed a model reference adaptive controller and obtained necessary conditions for achieving optimized performance with a uniform ultimate bounded solution using linear matrix inequalities (LMIs). The authors of (Yayla and Turker Kutay, 2016) proposed an indirect adaptive approach based on online identifications of matched and unmatched uncertainties, where the system performance may not be acceptable with this approach due to a modification to the reference model trajectories.

This chapter focuses on model reference adaptive control of dynamical systems with matched system uncertainties but unmatched disturbances. Departing from the above results, we propose a new, two-level design framework based on a command governor architecture to suppress the effect of matched uncertainties and unmatched disturbances and achieve a close tracking of the output of the reference system. Specifically, we first design an auxiliary state dynamics that allows not only the estimation of the matched uncertainties but also the estimation of the unmatched disturbances. Then, we propose a command governor architecture through a backstepping procedure to modify the command signal of the desired reference system such that the system output error signal can be made arbitrarily small by tuning the constant design parameters. Two numerical examples are provided to demonstrate the efficacy of the proposed command governor-based adaptive control architecture.

The organization of this chapter is as follows. In Section 2, we present necessary mathematical preliminaries. We state the problem formulation in Section 3 and introduce an auxiliary state dynamics design in Section 4. Section 5 presents a new command governor-based adaptive control architecture and two numerical examples are provided in Section 6 to demonstrate the efficacy of the proposed approach. Finally, we present conclusions in Section 7.

2. MATHEMATICAL PRELIMINARIES

We first introduce the standard notation used in this chapter. $\mathbb{R}$ denotes the set of real numbers, $\mathbb{R}^n$ denotes the set of $n \times 1$ real column vectors, $\mathbb{R}^{n \times m}$ denotes the set of $n \times m$ real matrices, $\mathbb{R}_+$ ($\overline{\mathbb{R}}_+$) denotes the set of positive (non-negative-definite) real numbers, $\mathbb{D}^{n \times n}$ denotes the set of $n \times n$ real matrices with diagonal scalar entries, $0_{n \times n}$ denotes the $n \times n$ zero matrix, and "$\triangleq$" denotes equality by definition. In addition, we write $(\cdot)^T$ for the transpose operator, $(\cdot)^{-1}$ for the inverse operator, $\mathrm{tr}(\cdot)$ for the trace operator, $\|\cdot\|_2$ for the Euclidean norm, and $\|A\|_2 \triangleq \sqrt{\lambda_{\max}(A^T A)}$ for the induced 2-norm of the matrix $A \in \mathbb{R}^{n \times m}$.

Let $\psi : \mathbb{R}^n \to \mathbb{R}$ be a *continuously differentiable* and *convex* function, and be given by $\psi(\theta) \triangleq \dfrac{(\varepsilon_\theta + 1)\theta^T\theta - \theta^2_{\max}}{\varepsilon_\theta \theta^2_{\max}}$, where $\theta_{\max} \in \mathbb{R}$ is a projection norm bound imposed on $\theta \in \mathbb{R}^n$ and $\varepsilon_\theta > 0$ is a projection tolerance bound. Then, the projection operator $\mathrm{Proj} : \mathbb{R}^n \times \mathbb{R}^n \to \mathbb{R}^n$ is defined by:

$$\mathrm{Proj}(\theta, y) \triangleq \begin{cases} y, & \text{if } \psi(\theta) < 0, \\ y, & \text{if } \psi(\theta) \geq 0 \text{ and } \psi'(\theta)\, y \leq 0, \\ y - \dfrac{\psi'^T(\theta)\, \psi'(\theta) y}{\psi'(\theta)\psi'^T(\theta)}\, \psi(\theta), & \text{if } \psi(\theta) \geq 0 \text{ and } \psi'(\theta)\, y > 0, \end{cases} \quad (1)$$

where $y \in \mathbb{R}^n$. It then follows that,

$$(\theta - \theta^*)^T \left[\mathrm{Proj}(\theta, y) - y \right] \leq 0, \ \theta^* \in \mathbb{R}^n, \quad (2)$$

holds (Pomet and Praly, 1992). The definition of the projection operator can be generalized to matrices as:

$$\text{Proj}_m(\Theta, Y) = (\text{Proj}(\text{col}_1(\Theta), \text{col}_1(Y)), \ldots, \text{Proj}(\text{col}_m(\Theta), \text{col}_m(Y))), \quad (3)$$

where $\Theta \in \mathbb{R}^{n \times m}$, $Y \in \mathbb{R}^{n \times m}$, and $\text{col}_i(\cdot)$ denotes i th column operator. In this case, for a given $\Theta^* \in \mathbb{R}^{n \times m}$, it follows from (2) that:

$$\text{tr}\left[(\Theta - \Theta^*)^\top[\text{Proj}_m(\Theta, Y) - Y]\right] = \sum_{i=1}^{m}\left[\text{col}_i(\Theta - \Theta^*)^\top\left[\text{Proj}(\text{col}_i(\Theta), \text{col}_i(Y)) - \text{col}_i(Y)\right]\right] \le 0. \quad (4)$$

Throughout this chapter, we assume without loss of generality that the projection norm bound imposed on each column of $\Theta \in \mathbb{R}^{n \times m}$ is θ_{max} through the *continuously differentiable* and *convex* function $\psi(\theta)$ defined above.

3. PROBLEM FORMULATION

We now introduce the problem considered throughout this chapter. For this purpose, consider the nonlinear uncertain dynamical system given by:

$$\dot{x}(t) = Ax(t) + B(\Lambda u(t) + \delta(t, x(t))) + Dq(t), \quad x(0) = x_0, \quad t \ge 0, \quad (5)$$

where $x(t) \in \mathbb{R}^n$, $t \ge 0$, is the measurable state vector, $u(t) \in \mathbb{R}^m$, $t \ge 0$, is the control input, $A \in \mathbb{R}^{n \times n}$ is a known system matrix, $B \in \mathbb{R}^{n \times m}$ is a known input matrix, $\delta : \mathbb{R}_+ \times \mathbb{R}^n \to \mathbb{R}^m$ is a system uncertainty, $\Lambda \in \mathbb{R}_+^{m \times m} \cap \mathbb{D}^{m \times m}$ is an unknown control effectiveness matrix, $D \in \mathbb{R}^{n \times (n-m)}$ is the unmatched disturbance input matrix such that $D^\top B = 0$ and rank$([B, D]) = n$ (this condition can be satisfied even when D has columns less than $(n-m)$ that is further discussed in Example 2 of Section 6), $q(t) \in \mathbb{R}^{(n-m)}$ is a bounded unmatched disturbance vector (i.e., $\| q(t) \|_2 \le \bar{q}, t \ge 0$) with a bounded time rate of change (i.e., $\| \dot{q}(t) \|_2 \le \bar{\dot{q}}, t \ge 0$), and the pair (A, B) is controllable. We now introduce a standard assumption on system uncertainty parameterization (Narendra and Annaswamy, 2012; Ioannou and Sun, 2012; Lavretsky and Wise, 2012).

Assumption 1. The system uncertainty given by (5) is parameterized as:

$$\delta(t, x(t)) = W_0^\top(t)\sigma_0(x(t)), \quad (6)$$

where $W_0(t) \in \mathbb{R}^{s \times m}$, $t \ge 0$, is a bounded unknown weight matrix (i.e., $\| W_0(t) \|_2 \le w_0, t \ge 0$) with a bounded time rate of change (i.e., $\| \dot{W}_0(t) \|_2 \le \dot{w}_0, t \ge 0$) and $\sigma_0 : \mathbb{R}^n \to \mathbb{R}^s$ is a known basis function of the form $\sigma_0(x(t)) = [\sigma_{01}(x(t)), \sigma_{02}(x(t)), \ldots, \sigma_{0s}(x(t))]^\top$.

Now, we consider the ideal reference model dynamics which captures a desired closed-loop dynamical system performance and is given by:

$$\dot{x}_r(t) = A_r x_r(t) + B_r c_d(t), \quad x_r(0) = x_{r0}, \quad t \geq 0, \tag{7}$$

where $x_r(t) \in \mathbb{R}^n$, $t \geq 0$, is the reference state vector, $c_d(t) \in \mathbb{R}^{nc}$ is the desired uniformly continuous bounded command, $A_r \in \mathbb{R}^{n \times n}$ is the Hurwitz reference model matrix, and $B_r \in \mathbb{R}^{n \times nc}$ is the command input matrix.

In this chapter, our goal is to drive a selected subset of system states given by:

$$y(t) = Cx(t), \quad t \geq 0, \tag{8}$$

to a close neighborhood of the selected subset of the reference system states given by:

$$y_r(t) = Cx_r(t), \quad t \geq 0. \tag{9}$$

For this purpose, the control design is presented in two sections. In Section 4, an auxiliary state is introduced to allow not only the estimation of the matched uncertainties but also the estimation of the unmatched disturbances and in Section 5 we present a command governor-based approach for achieving close tracking of selected system error states.

4. AUXILIARY STATE DYNAMICS AND ADAPTIVE CONTROL LAWS

For the nonlinear uncertain dynamical system introduced in the previous section, we now introduce an auxiliary state dynamics in order to analyze the effect of the unmatched system disturbances on the system. To this end, using Assumption 1, one can rewrite (5) as:

$$\dot{x}(t) = Ax(t) + B\left(\Lambda u(t) + W_0^T(t)\sigma_0(x(t))\right) + Dq(t), \quad x(0) = x_0, \quad t \geq 0. \tag{10}$$

Consider now, by adding and subtracting the terms $BK_1 x(t)$, $t \geq 0$ and $BK_2 c(t)$, $t \geq 0$, the following equivalent form of (10):

$$\dot{x}(t) = A_r x(t) + B_r c(t) + B\Lambda(u(t) + W^T(t)\sigma(x(t), c(t))) + Dq(t), \quad x(0) = x_0, \quad t \geq 0, \tag{11}$$

where $A_r \triangleq A - BK_1$, $B_r \triangleq BK_2$, $K_1 \in \mathbb{R}^{m \times n}$ is a feedback gain matrix, $K_2 \in \mathbb{R}^{m \times nc}$ is a feedforward gain matrix, $W(t) \triangleq [\Lambda^{-1} W_0^T(t), \Lambda^{-1} K_1, -\Lambda^{-1} K_2]^T \in \mathbb{R}^{(s+n+nc) \times m}$, $t \geq 0$, is an unknown aggregated weight matrix, and $\sigma(x(t), c(t)) \triangleq [\sigma_0^T(x(t)), x^T(t), c^T(t)]^T \in \mathbb{R}^{s+n+nc}$, $t \geq 0$, is a known aggregated basis function. Note that $\| W(t) \|_2 \leq w$, $t \geq 0$, and $\| \dot{W}(t) \|_2 \leq \dot{w}$, $t \geq 0$, automatically holds as a direct consequence of Assumption 1. In addition, we consider $c(t) \in \mathbb{R}^{n_c}$ to be the actual applied command signal given by:

$$c(t) \triangleq c_d(t) + c_g(t), \quad t \geq 0, \tag{12}$$

with $c_g(t) \in \mathbb{R}^{n_c}$, $t \geq 0$, being a modification term to the ideal command $c_d(t)$, $t \geq 0$. This modification term will be designed in Section 5.

We now define the auxiliary state dynamics as:

$$\dot{x}_a(t) = A_r x_a(t) + B_r c(t) + D\hat{q}(t), \quad x_a(0) = x_{a0}, \quad t \geq 0, \tag{13}$$

$$y_a(t) = C x_a(t), \quad t \geq 0, \tag{14}$$

where $x_a(t) \in \mathbb{R}^n$, $t \geq 0$, is the auxiliary state, $y_a(t) \in \mathbb{R}^{n_y}$, $t \geq 0$, is the auxiliary output signal, and $\hat{q}(t) \in \mathbb{R}^{(n-m)}$, $t \geq 0$, is the estimation of the unmatched disturbance $q(t)$, $t \geq 0$, satisfying the update law:

$$\dot{\hat{q}}(t) = \gamma_q \, \mathrm{Proj}(\hat{q}(t), D^\mathrm{T} P e_a(t)), \quad \hat{q}(0) = \hat{q}_0, \quad t \geq 0, \tag{15}$$

with $\hat{q}_{max}$ being the projection norm bound, and $e_a(t) \triangleq x(t) - x_a(t)$, $t \geq 0$, being the auxiliary error. Considering (11), let the adaptive control law be given by:

$$u(t) = -\hat{W}^\mathrm{T}(t)\sigma(x(t), c(t)), \quad t \geq 0, \tag{16}$$

where $\hat{W}(t) \in \mathbb{R}^{(s+n+n_c) \times m}$, $t \geq 0$, is the estimate of $W(t)$, $t \geq 0$, satisfying the update law:

$$\dot{\hat{W}}(t) = \gamma_W \, \mathrm{Proj}_m(\hat{W}(t), \sigma(x(t), c(t)) e_a^\mathrm{T}(t) PB), \quad \hat{W}(0) = \hat{W}_0, \quad t \geq 0, \tag{17}$$

with $\hat{W}_{max}$ being the projection norm bound. In (15) and (17), $\gamma_W, \gamma_q \in R_+$ and are the learning rates (adaptation gains), and $P \in \mathbb{R}_+^{n \times n}$ is a solution of the Lyapunov equation given by:

$$0 = A_r^\mathrm{T} P + P A_r + R, \tag{18}$$

with $R \in \mathbb{R}_+^{n \times n}$.

Using (11), (13), and (16), the auxiliary error dynamics can be written as:

$$\dot{e}_a(t) = A_r e_a(t) - B \Lambda \tilde{W}^\mathrm{T}(t)\sigma(x(t), c(t)) - D\tilde{q}(t), \quad e_a(0) = e_{a0}, \quad t \geq 0, \tag{19}$$

where $\tilde{W}(t) \triangleq \hat{W}(t) - W(t) \in \mathbb{R}^{(s+n+n_c) \times m}$, $t \geq 0$, is the weight estimation error, and $\tilde{q}(t) \triangleq \hat{q}(t) - q(t) \in \mathbb{R}^{(n-m)}$, $t \geq 0$, is the unmatched disturbance estimation error. Furthermore, using (15) and (17), one can write the unmatched disturbance estimation error and the weight estimation error dynamics as:

$$\dot{\tilde{q}}(t) = \gamma_q \operatorname{Proj}(\hat{q}(t), D^{\mathrm{T}}Pe_a(t)) - \dot{q}(t), \quad \tilde{q}(0) = \tilde{q}_0, \quad t \geq 0, \tag{20}$$

$$\dot{\tilde{W}}(t) = \gamma_W \operatorname{Proj}_m(\hat{W}(t), \sigma(x(t), c(t))e_a^{\mathrm{T}}(t)PB) - \dot{W}(t), \quad \tilde{W}(0) = \tilde{W}_0, \quad t \geq 0. \tag{21}$$

Theorem 1. Consider the uncertain dynamical system given by (5) subject to Assumption 1, the auxiliary state dynamics given by (13) and (14) along with the update law (15), and the feedback control law given by (16) along with the update law (17), then the closed-loop dynamical system given by (19), (20), and (21) are uniformly bounded.

Proof. To show boundedness of the closed-loop dynamical system given by (19), (20), and (21), consider the Lyapunov function candidate $V : \mathbb{R}^n \times \mathbb{R}^{(s+n+n_c)\times m} \times \mathbb{R}^{(n-m)} \to \mathbb{R}_+$ given by:

$$V(e_a, \tilde{W}, \tilde{q}) = e_a^{\mathrm{T}}Pe_a + \gamma_W^{-1}\operatorname{tr}[(\tilde{W}\Lambda^{1/2})^{\mathrm{T}}(\tilde{W}\Lambda^{1/2})] + \gamma_q^{-1}\tilde{q}^{\mathrm{T}}\tilde{q}. \tag{22}$$

where $P \in \mathbb{R}_+^{n\times n}$ is a solution of the Lyapunov equation in (18) with $R \in \mathbb{R}_+^{n\times n}$. Note that $V(0, 0, 0) = 0$, $V(e_a, \tilde{W}, \tilde{q}) > 0$ for all $(e, \tilde{W}, \tilde{q}) \neq (0, 0, 0)$. The time derivative of (22) along the closed-loop system trajectories (19), (20), and (21) is given by:

$$\begin{aligned}
\dot{V}(e_a(t), \tilde{W}(t), \tilde{q}(t)) &= 2e_a^{\mathrm{T}}(t)PA_re_a(t) - 2e_a^{\mathrm{T}}(t)PB\,\Lambda\tilde{W}^{\mathrm{T}}(t)\sigma(x(t), c(t)) - 2e_a^{\mathrm{T}}(t)PD\,\tilde{q}(t) \\
&\quad + 2\operatorname{tr}\tilde{W}^{\mathrm{T}}(t)\operatorname{Proj}_m(\hat{W}(t), \sigma(x(t), c(t))e_a^{\mathrm{T}}(t)PB) - 2\gamma_W^{-1}\operatorname{tr}\tilde{W}^{\mathrm{T}}(t)\dot{W}(t)\Lambda \\
&\quad + 2\tilde{q}^{\mathrm{T}}(t)\operatorname{Proj}(\hat{q}(t), D^{\mathrm{T}}Pe_a(t)) - 2\gamma_q^{-1}\tilde{q}^{\mathrm{T}}(t)\dot{q}(t) \\
&= -e_a^{\mathrm{T}}(t)Re_a(t) - 2\gamma_W^{-1}\operatorname{tr}\tilde{W}^{\mathrm{T}}(t)\dot{W}(t)\Lambda - 2\gamma_q^{-1}\tilde{q}^{\mathrm{T}}(t)\dot{q}(t) \\
&\quad + 2\operatorname{tr}\tilde{W}^{\mathrm{T}}(t)(\operatorname{Proj}_m(\hat{W}(t), \sigma(x(t), c(t))e_a^{\mathrm{T}}(t)PB) \\
&\quad - \sigma(x(t), c(t))e_a^{\mathrm{T}}(t)PB)\Lambda + 2\tilde{q}^{\mathrm{T}}(t)(\operatorname{Proj}(\hat{q}(t), D^{\mathrm{T}}Pe_a(t)B) - D^{\mathrm{T}}Pe_a(t)) \\
&\leq -\lambda_{\min}(R)\|e_a(t)\|_2^2 + 2\gamma_W^{-1}\tilde{w}\dot{w}\|\Lambda\|_2 + 2\gamma_q^{-1}\tilde{q}_0\dot{q}, \tag{23}
\end{aligned}$$

where $\tilde{w} = \hat{W}_{\max} + w$, $\tilde{q}_0 = \hat{q}_{\max} + \bar{q}$. Hence, $\dot{V}(e(t), \tilde{W}(t), \tilde{q}(t)) < 0$, $t \geq 0$, outside of the compact set:

$$\Omega \triangleq \left\{ (e_a(t), \tilde{W}(t), \tilde{q}(t)) : \|e_a(t)\|_2 \leq \eta, \ \|\tilde{W}(t)\|_2 \leq \tilde{w}, \text{ and } \|\tilde{q}(t)\|_2 \leq \tilde{q}_0 \right\}, \tag{24}$$

where $\eta \triangleq \sqrt{\dfrac{2\gamma_W^{-1}\tilde{w}\dot{w}\|\Lambda\|_2 + 2\gamma_q^{-1}\tilde{q}_0\dot{q}}{\lambda_{\min}(R)}}$. From (24), one can conclude uniform boundedness of the solution $(e_a(t), \tilde{W}(t), \tilde{q}(t))$ of the system dynamics given by (19), (20), and (21) for all $(e_{a0}, \tilde{W}_0, \tilde{q}_0) \in \mathbb{R}^n \times \mathbb{R}^{(s+n+n_c)\times m} \times \mathbb{R}^{(n-m)}$.

Remark 1. Note that one cannot conclude boundedness of the auxiliary state, $x_a(t), t \geq 0$, using the analysis in this section. Theorem 1 only guarantees that the auxiliary error signal $e_a(t)$, $t \geq 0$, is ultimately bounded. However, it is possible that the auxiliary state signal $x_a(t)$, $t \geq 0$, and the system state signal $x(t)$, $t \geq 0$, both grow to infinity such that their difference remains bounded. For the case where no modification is applied to the desired command signal (i.e., $c_g(t) \equiv 0$, $t \geq 0$), and also there are no unmatched disturbances in the system, the auxiliary state dynamics become:

$$\dot{x}_a(t) = A_r x_a(t) + B_r c_d(t), \quad x_a(0) = x_{a0}, \quad t \geq 0, \tag{25}$$

$$y_a(t) = C x_a(t), \quad t \geq 0. \tag{26}$$

The auxiliary state dynamics given by (25) and (26) exactly capture the ideal reference system behavior given by (7) and (9). In this special case, the auxiliary state signal $x_a(t)$, $t \geq 0$, is bounded, therefore the system state signal $x(t)$, $t \geq 0$, will be ultimately bounded as well. For the case where the unmatched disturbances are present, the next section applies a command governor signal to the desired command signal (i.e., $c_g(t) \not\equiv 0$, $t \geq 0$) such that a selected subset of the auxiliary state signal $y_a(t)$, $t \geq 0$, can be kept within a close and adjustable neighborhood of the reference system output $y_r(t)$, $t \geq 0$, which makes the proposed control architecture go beyond the results presented in (Yayla and Turker Kutay, 2016).

5. DESIGN OF THE COMMAND GOVERNOR

In this section, we introduce and analyze a novel command governor-based adaptive control architecture to suppress the effect of matched system uncertainties and unmatched system disturbances. For this purpose, let $e_r(t) \triangleq x_a(t) - x_r(t)$, $t \geq 0$, be the error signal between the auxiliary state and the ideal reference state, and let the modification term of the command signal be given by:

$$c_g(t) \triangleq K_2^{-1} \xi(t), \quad t \geq 0, \tag{27}$$

where $\xi(t) \in \mathbb{R}^m$, $t \geq 0$, is the command governor signal to be designed. One can write the system error dynamics between the auxiliary dynamics and the ideal reference system as:

$$\dot{e}_r(t) = A_r e_r(t) + B\xi(t) + D\hat{q}(t), \quad e_r(0) = x_{a0} - x_{r0}, \quad t \geq 0, \tag{28}$$

$$e_y(t) = C e_r(t), \quad t \geq 0, \tag{29}$$

where $e_y(t) \triangleq y_a(t) - y_r(t)$, $t \geq 0$.

In what follows, we systematically show that one can employ the backstepping control methodology to design the command governor signal $\xi(t)$, $t \geq 0$, in (28) to guarantee the boundedness of the error signal

$e_r(t)$, $t \geq 0$. In addition, the norm of the output error signal $e_y(t)$, $t \geq 0$, can be made arbitrarily small as desired. For this purpose, we consider (28) and (29) in the control canonical form with:

$$A_r = \begin{bmatrix} 0 & 1 & 0 & \cdots & 0 \\ 0 & 0 & 1 & \cdots & 0 \\ \vdots & \vdots & \vdots & \ddots & \vdots \\ 0 & 0 & 0 & \cdots & 1 \\ -k_1 & -k_2 & -k_3 & \cdots & -k_n \end{bmatrix}, \quad k_i \in \mathbb{R}, \tag{30}$$

$$B = \begin{bmatrix} 0_{(\rho-1)\times m} \\ \overline{B} \end{bmatrix}, \quad \overline{B} = \begin{bmatrix} b_{\rho 1} & \cdots & b_{\rho m} \\ \vdots & \ddots & \vdots \\ b_{n1} & \cdots & b_{nm} \end{bmatrix} \in \mathbb{R}^{(n-\rho+1)\times m}, \quad b_i \in \mathbb{R}, \tag{31}$$

$$C = \begin{bmatrix} 1 & 0 & \cdots & 0 \end{bmatrix},$$

$$D = \begin{bmatrix} 0 & \overline{D} \\ & 0_{(n-\rho+1)\times(n-m)} \end{bmatrix}, \quad \overline{D} = \begin{bmatrix} d_{11} & \cdots & d_{1(n-m)} \\ \vdots & \ddots & \vdots \\ d_{(\rho-1)1} & \cdots & d_{(\rho-1)(n-m)} \end{bmatrix} \in \mathbb{R}^{(\rho-1)\times(n-m)}, \quad d_i \in \mathbb{R}. \tag{33}$$

Now, let $\hat{q}_f(t) \in \mathbb{R}^{(n-m)}$, $t \geq 0$, be a low-pass filter estimate of $\hat{q}(t)$, $t \geq 0$, given by:

$$\dot{\hat{q}}_f(t) = \Gamma_f[\hat{q}(t) - \hat{q}_f(t)], \quad \hat{q}_f(0) = \hat{q}_{f0}, \quad t \geq 0, \tag{34}$$

where $\Gamma_f \in \mathrm{ID}^{(n-m)\times(n-m)}$ is a positive-definite filter gain matrix. Note that since $\hat{q}_f(t)$, $t \geq 0$, is a low-pass filter estimate of $\hat{q}(t)$, $t \geq 0$, the filter gain matrix Γ_f is chosen such that $\lambda_{max}(\Gamma_f) \leq \gamma_{f,max}$, with $\gamma_{f,max} > 0$ being a design parameter.

Remark 2. Note that since $\hat{q}(t)$, $t \geq 0$, is a bounded signal and the filter gain matrix Γ_f is positive-definite, it follows from (34) that $\hat{q}_f(t)$, $t \geq 0$ and $\dot{\hat{q}}_f(t)$, $t \geq 0$ are bounded.

Next, in order to obtain a recursive procedure using a backstepping control design, as standard, we start with the second-order system given by:

$$\dot{e}_{r1}(t) = e_{r2}(t) + d_1 \hat{q}_1(t), \quad e_{r1}(0) = e_{r10}, \quad t \geq 0, \tag{35}$$

$$\dot{e}_{r2}(t) = -k_1 e_{r1}(t) - k_2 e_{r2}(t) + b_1 \xi(t), \quad e_{r2}(0) = e_{r20}, \quad t \geq 0, \tag{36}$$

$$e_y(t) = e_{r1}(t), \quad t \geq 0. \tag{37}$$

Letting $\varepsilon_1(t) \triangleq \Gamma_0 e_{r1}(t) + e_{r2}(t) + d_1 \hat{q}_{1f}(t)$, $t \geq 0$, we design the command governor signal as:

$$\zeta(t) \triangleq -b_1^{-1}\Big[(\Gamma_1 + \Gamma_0 - k_2)\varepsilon_1(t) - (\Gamma_0^2 - k_2\Gamma_0 + k_1)e_{r1}(t) + \Gamma_0 d_1(\hat{q}_1(t) - \hat{q}_{1f}(t))$$

$$+ k_2 d_1 \hat{q}_{1f}(t) + d_1 \dot{\hat{q}}_{1f}(t)\Big], \quad t \geq 0, \tag{38}$$

where $\Gamma_0, \Gamma_1 \in \mathbb{R}$ are design parameters. Using the new state variable $\varepsilon_1(t)$, $t \geq 0$, and the command governor signal given by (38), the system error dynamics in (35), (36), and (37) can be rewritten as:

$$\dot{e}_{r1}(t) = -\Gamma_0 e_{r1}(t) + \varepsilon_1(t) + d_1(\hat{q}_1(t) - \hat{q}_{1f}(t)), \; e_{r1}(0) = e_{r1_0}, \quad t \geq 0, \tag{39}$$

$$\dot{\varepsilon}_1(t) = -\Gamma_1 \varepsilon_1(t), \quad \varepsilon_1(0) = \varepsilon_{1_0}, \quad t \geq 0, \tag{40}$$

$$e_y(t) = e_{r1}(t), \quad t \geq 0, \tag{41}$$

which can be written in compact form as:

$$\dot{\zeta}(t) = A_1 \zeta(t) + B_1 \tilde{q}_1(t), \quad \zeta(0) = \zeta_0, \quad t \geq 0, \tag{42}$$

$$e_y(t) = C_1 \zeta(t), \quad t \geq 0, \tag{43}$$

with,

$$A_1 = \begin{bmatrix} -\Gamma_0 & 1 \\ 0 & -\Gamma_1 \end{bmatrix}, \quad B_1 = \begin{bmatrix} 1 \\ 0 \end{bmatrix}, \quad C_1 = \begin{bmatrix} 1 & 0 \end{bmatrix}, \tag{44}$$

where $\zeta(t) = [e_{r1}(t), \varepsilon_1(t)]^T$, $t \geq 0$, is the aggregated system state, and $\tilde{q}_1(t) \triangleq d_1(\hat{q}_1(t) - \hat{q}_{1f}(t))$, $t \geq 0$, is a bounded signal as noted in Remark 2. Therefore, it follows from (Haddad and Chellaboina, 2008) that $e_{r1}(t)$, $t \geq 0$, and $\varepsilon_1(t)$, $t \geq 0$, are bounded, and hence, the error signal $e_r(t)$, $t \geq 0$, is bounded which results in the boundedness of the auxiliary state $x_a(t)$, $t \geq 0$.

From a practical point of view, we are interested in analyzing how small is the output error signal in (43). Therefore, we write the $\mathcal{L}_1$-system norm of (42) and (43) by the equi-induced signal norm (Yucelen and Haddad, 2012; Chellaboina et al., 2000) as:

$$|||\mathcal{G}|||_{(\infty,2),(\infty,2)} \triangleq \sup_{\tilde{q}_1 \in \mathcal{L}_\infty} \frac{|||e_y|||_{\infty,2}}{|||\tilde{q}_1|||_{\infty,2}}, \tag{45}$$

which has an upper bound given by:

$$|||\mathcal{G}|||_{(\infty,2),(\infty,2)} \leq \frac{1}{\sqrt{\alpha}} \sigma_{max}^{1/2}(C_1 Q_\alpha C_1^T), \tag{46}$$

where $\alpha > 0$ is selected such that $A_1 + \dfrac{\alpha}{2}I$ is Hurwitz, and $Q_\alpha \in \mathbb{R}^{2 \times 2}$ is the unique, non-negative definite solution to the Lyapunov equation

$$0 = A_1 Q_\alpha + Q_\alpha A_1^T + \alpha Q_\alpha + B_1 B_1^T. \tag{47}$$

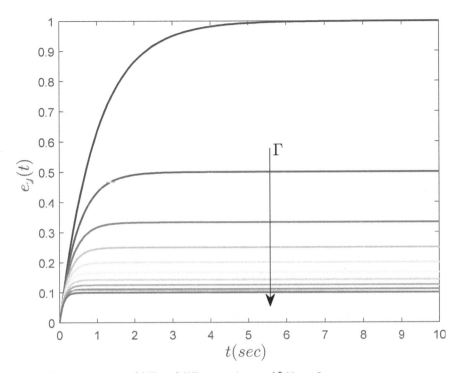

Fig. 1. System response of (42) and (43) to step input of $\tilde{q}_1(t)$, $t \geq 0$.

Remark 3. For the purpose of understanding the ability of the design parameters Γ_0 and Γ_1 in (42) and (43) to suppress the effect of $\tilde{q}_1(t)$, $t \geq 0$, we consider $\tilde{q}_1(t)$, $t \geq 0$, to be a unit step input. As depicted in Fig. 1, $e_y(t)$, $t \geq 0$, decreases as $\Gamma \triangleq \Gamma_0 = \Gamma_1$ increases from 1 to 10. Furthermore, the upper bound of the $\mathcal{L}_1$-system norm of (42) and (43) is shown in Fig. 2, where the $\mathcal{L}_1$-system norm can be made arbitrarily small by increasing the design parameter Γ.

The same procedure can be recursively employed due to the nature of the backstepping approach to obtain the command governor signal $\xi(t)$, $t \geq 0$, for the high-order dynamical systems to guarantee the boundedness of the auxiliary state signal $x_a(t)$, $t \geq 0$ and to make the output of the auxiliary dynamics arbitrarily close to the output of the reference system by tuning the design parameters. To elucidate this point, consider the third order system given by:

$$\dot{e}_{r1}(t) = e_{r2}(t) + d_{11}\hat{q}_1(t) + d_{12}\hat{q}_2(t), \qquad e_{r1}(0) = e_{r10}, \quad t \geq 0, \qquad (48)$$

$$\dot{e}_{r2}(t) = e_{r3}(t) + d_{21}\hat{q}_1(t) + d_{22}\hat{q}_2(t), \qquad e_{r2}(0) = e_{r20}, \quad t \geq 0, \qquad (49)$$

$$\dot{e}_{r3}(t) = -k_1 e_{r1}(t) - k_2 e_{r2}(t) - k_3 e_{r3}(t) + b_1 \xi(t), \quad e_{r3}(0) = e_{r30}, \quad t \geq 0, \qquad (50)$$

$$e_y(t) = e_{r1}(t), \quad t \geq 0. \qquad (51)$$

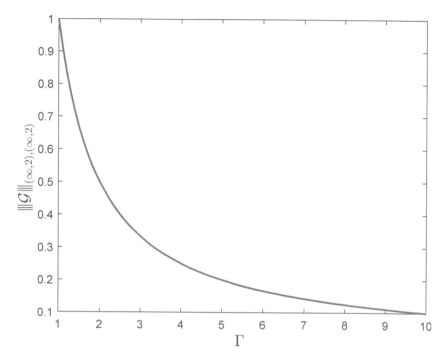

Fig. 2. The upper bound on the $\mathcal{L}_1$-system norm of (42) and (43) given in (46).

Now, we let,

$$\varepsilon_1(t) \triangleq \Gamma_0 e_{r1}(t) + e_{r2}(t) + d_{11}\hat{q}_{1f}(t) + d_{12}\hat{q}_{2f}(t), \quad t \geq 0, \tag{52}$$

$$\varepsilon_2(t) \triangleq (\Gamma_1 + \Gamma_0)\varepsilon_1(t) - \Gamma_0^2 e_{r1}(t) + e_{r3}(t) + d_{21}\hat{q}_{1f}(t) + d_{22}\hat{q}_{2f}(t)$$
$$+ d_{11}\dot{\hat{q}}_{1f}(t) + d_{12}\dot{\hat{q}}_{2f}(t), \quad t \geq 0, \tag{53}$$

and we design the command governor signal as:

$$\zeta(t) \triangleq -b_1^{-1}\Big[(\Gamma_2 + \Gamma_1 + \Gamma_0 - k_3)\varepsilon_2(t) - (\Gamma_1^2 + \Gamma_1\Gamma_0 + \Gamma_0^2 - k_3(\Gamma_1 + \Gamma_0) - k_2)\varepsilon_1(t) + (\Gamma_0^3 - k_3\Gamma_0^2$$
$$+ k_2\Gamma_0 - k_1)e_{r1}(t) + \Gamma_1\Gamma_0(d_{11}(\hat{q}_1(t) - \hat{q}_{1f}(t)) + d_{12}(\hat{q}_2(t) - \hat{q}_{2f}(t))) + (\Gamma_1 + \Gamma_0)$$
$$\cdot(d_{21}(\hat{q}_1(t) - \hat{q}_{1f}(t)) + d_{22}(\hat{q}_2(t) - \hat{q}_{2f}(t))) + k_2(d_{11}\hat{q}_{1f}(t) + d_{12}\hat{q}_{2f}(t)) + k_3(d_{21}\hat{q}_{1f}(t)$$
$$+ d_{22}\hat{q}_{2f}(t)) + (d_{11}k_3 + d_{21})\dot{\hat{q}}_{1f}(t) + (d_{12}k_3 + d_{22})\dot{\hat{q}}_{2f}(t) + d_{11}\ddot{\hat{q}}_{1f}(t) + d_{12}\ddot{\hat{q}}_{2f}(t)\Big], \quad t \geq 0, \tag{54}$$

where $\Gamma_0, \Gamma_1, \Gamma_2 \in \mathbb{R}$ are design parameters. Using the new state variables $\varepsilon_1(t), t \geq 0, \varepsilon_2(t), t \geq 0$, and the command governor signal given by (54), the system error dynamics in (48), (49), (50), and (51) can be rewritten as:

$$\dot{e}_{r1}(t) = -\Gamma_0 e_{r1}(t) + \varepsilon_1(t) + d_{11}(\hat{q}_1(t) - \hat{q}_{1f}(t)) + d_{12}(\hat{q}_2(t) - \hat{q}_{2f}(t)), \quad e_{r1}(0) = e_{r10}, \quad t \geq 0, \tag{55}$$

$$\dot{\varepsilon}_1(t) = -\Gamma_1\varepsilon_1(t) + \varepsilon_2(t) + \Gamma_0\left[d_{11}(\hat{q}_1(t) - \hat{q}_{1f}(t)) + d_{12}(\hat{q}_2(t) - \hat{q}_{2f}(t))\right] + d_{21}(\hat{q}_1(t) - \hat{q}_{1f}(t))$$
$$+ d_{22}(\hat{q}_2(t) - \hat{q}_{2f}(t)), \quad \varepsilon_1(0) = \varepsilon_{10}, \quad t \geq 0, \tag{56}$$

$$\dot{\varepsilon}_2(t) = -\Gamma_2\varepsilon_2(t), \quad \varepsilon_2(0) = \varepsilon_{20}, \quad t \geq 0, \tag{57}$$

$$e_y(t) = e_{r1}(t), \quad t \geq 0, \tag{58}$$

which can be written in compact form as:

$$\dot{\zeta}(t) = A_2\zeta(t) + B_2\tilde{q}_2(t), \quad \zeta(0) = \zeta_0, \quad t \geq 0, \tag{59}$$

$$e_y(t) = C_2\zeta(t), \quad t \geq 0, \tag{60}$$

with,

$$A_2 = \begin{bmatrix} -\Gamma_0 & 1 & 0 \\ 0 & -\Gamma_1 & 1 \\ 0 & 0 & -\Gamma_2 \end{bmatrix}, \quad B_2 = \begin{bmatrix} 1 & 0 \\ \Gamma_0 & 1 \\ 0 & 0 \end{bmatrix}, \quad C_2 = \begin{bmatrix} 1 & 0 & 0 \end{bmatrix}, \tag{61}$$

where $\zeta(t) = [e_{r1}(t), \varepsilon_1(t), \varepsilon_2(t)]^T$, $t \geq 0$ is the aggregated system state, and,

$$\tilde{q}_2(t) - \begin{bmatrix} d_{11}(\hat{q}_1(t) - \hat{q}_{1f}(t)) + d_{12}(\hat{q}_2(t) - \hat{q}_{2f}(t)) \\ d_{21}(\hat{q}_1(t) - \hat{q}_{1f}(t)) + d_{22}(\hat{q}_2(t) - \hat{q}_{2f}(t)) \end{bmatrix}, \quad t \geq 0, \tag{62}$$

is a bounded signal as noted in Remark 2. Similar to the previous case, it follows from (Haddad and Chellaboina, 2008) that $e_{r1}(t)$, $t \geq 0$, $\varepsilon_1(t)$, $t \geq 0$, and $\varepsilon_2(t)$, $t \geq 0$, are bounded, and hence, the error signal $e_r(t)$, $t \geq 0$, is bounded which results in the boundedness of the state auxiliary state $x_a(t)$, $t \geq 0$.

Similar to (45), one can write $\mathcal{L}_1$-system norm of (59) and (60) as:

$$|||\mathcal{G}|||_{(\infty,2),(\infty,2)} \triangleq \sup_{\tilde{q}_2 \in \mathcal{L}_\infty} \frac{|||e_y|||_{\infty,2}}{|||\tilde{q}_2|||_{\infty,2}} \tag{63}$$

where it follows from (Yucelen and Haddad, 2012; Chellaboina et al., 2000) that:

$$|||\mathcal{G}|||_{(\infty,2),(\infty,2)} \leq \frac{1}{\sqrt{\alpha}}\sigma_{\max}^{1/2}(C_2 Q_\alpha C_2^T), \tag{64}$$

where $\alpha > 0$ is selected such that $A_2 + \frac{\alpha}{2}I$ is Hurwitz, and $Q_\alpha \in \mathbb{R}^{3\times3}$ is the unique, non-negative definite solution to the Lyapunov equation:

$$0 = A_2 Q_\alpha + Q_\alpha A_2^T + \alpha Q_\alpha + B_2 B_2^T. \tag{65}$$

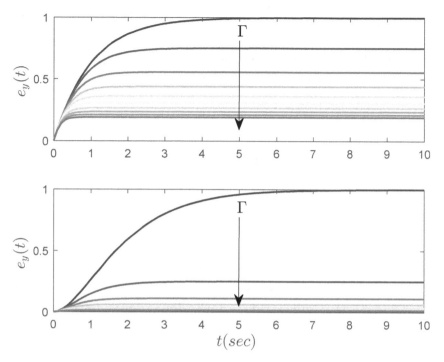

Fig. 3. System response of (59) and (60) to step input of the first (top) and the second (bottom) components of $\tilde{q}_2(t)$, $t \geq 0$, in (62).

Remark 4. As depicted in Fig. 3, $e_y(t)$, $t \geq 0$, decreases as $\Gamma \triangleq \Gamma_0 = \Gamma_1 = \Gamma_2$ increases from 1 to 10, similar to the case in Remark 3. The upper bound of the $\mathcal{L}_1$-system norm of (59) and (60) is shown in Fig. 4, where one can decrease the $\mathcal{L}_1$-system norm by increasing the design parameter Γ.

Repeating the recursive procedure outlined above in this section $(n-1)$-times, the command governor signal $\xi(t)$, $t \geq 0$, can be obtained for the general error dynamical systems given in (28) and (29) to guarantee the boundedness of the auxiliary state signal $x_a(t)$, $t \geq 0$ and to tighten the upper bound on the output error signal in (29) by tuning the design parameters Γ_0, Γ_1,... Γ_{n-1}.

Remark 5. It is worth noting that the selection of the canonical structure of the reference system in (30) to (33) is without loss of generality. Specifically, for different structures for the reference system, one can perform a similar recursive backstepping approach in order to design the command governor signal $\xi(t)$, $t \geq 0$.

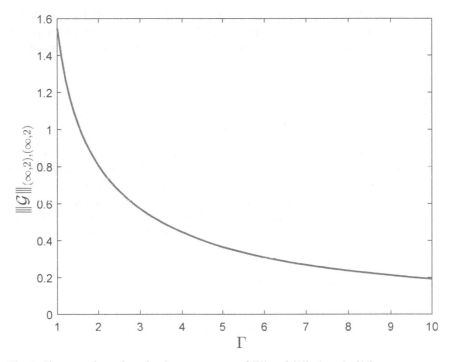

Fig. 4. The upper bound on the $\mathcal{L}_1$-system norm of (59) and (60) given in (64).

6. ILLUSTRATIVE NUMERICAL EXAMPLE

In this section, we present two numerical examples to demonstrate the efficacy of the proposed command governor-based adaptive control architecture.

Example 1. Consider the uncertain dynamical system given by:

$$\dot{x}(t) = \begin{bmatrix} 0 & 1 \\ 2 & 4 \end{bmatrix} x(t) + \begin{bmatrix} 0 \\ 1 \end{bmatrix} \Big(\Lambda u(t) + \delta(t, x(t)) \Big) + \begin{bmatrix} 1 \\ 0 \end{bmatrix} q(t), \quad x(0) = 0, \quad t \geq 0, \quad (66)$$

where $x(t) = [x_1(t) \; x_2(t)]^T$ is the system state, $\delta(t, x(t))$ represents an uncertainty of the form given in (6) with:

$$W_0(t) = [\sin(0.25t), -1, 1]^T, \quad t \geq 0, \quad \sigma_0(x(t)) = [x_1(t), x_2(t), x_1(t)x_2(t)]^T, \quad t \geq 0, \quad (67)$$

$q(t) = 0.5 \sin(0.2t)$, $t \geq 0$, represents the unmatched disturbance, and $\Lambda = 0.75$ represents an uncertain control effectiveness matrix. Linear quadratic regulator theory is used to design the nominal feedback gain matrix as:

$$K_1 = [5.7, 9.7], \tag{68}$$

and we pick $K_2 = 3.7$.

For the adaptive controller in Section 4, we set the projection norm bound imposed on each element of the parameter estimate to $\hat{W}_{max} = 30$ and $\hat{q}_{max} = 5$ and the learning rates to $\gamma_q = 5$ and $\gamma_w = 30$ and we use $R = I$ to calculate P from (18) for the resulting A_r matrix. Fig. 5 shows the closed-loop dynamical system performance with the standard adaptive controller in Section 4. One can see from this figure that the standard adaptive controller cannot compensate the effects of the unmatched disturbance and the system's trajectories do not converge to the reference system trajectory.

Next, we apply the proposed command governor-based adaptive control architecture. For this purpose, we use $\Gamma_0 = \Gamma_1 = 10$ and set the filter gain in (34) to $\Gamma_f = 0.5$. It can be seen in Fig. 6 that desired performance is obtained and the first component of the state vector converges to a close vicinity of the reference state. Fig. 7 clearly shows the role of the

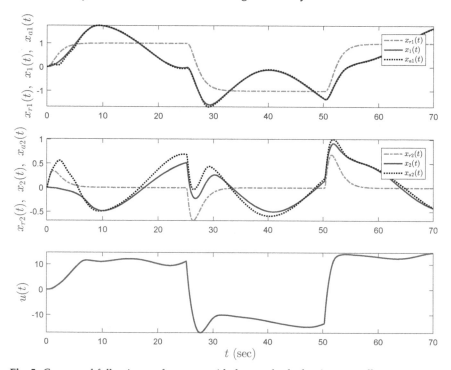

Fig. 5. Command following performance with the standard adaptive controller.

command governor signal to modify the command signal such that the first component of the error signal $e_{r1}(t)$, $t \geq 0$, gets arbitrarily close to zero by tuning the design gains Γ_0 and Γ_1 as one can see in Fig. 8. The evolution of the unmatched disturbance estimation is depicted in Fig. 9. Finally, the effect of the design parameter $\Gamma = \Gamma_0 = \Gamma_1$ can be seen in Figs. 10 and 11 where it is clear that a larger value of Γ, leads to a better tracking performance of output signal of the reference system.

Example 2. For this second example, we consider a third-order uncertain dynamical system given by:

$$\dot{x}(t) = \begin{bmatrix} 0 & 1 & 0 \\ 0 & 0 & 1 \\ 2 & 3 & 1 \end{bmatrix} x(t) + \begin{bmatrix} 0 \\ 0 \\ 1 \end{bmatrix} \left(\Lambda u(t) + \delta(t, x(t)) \right) + \begin{bmatrix} 1 \\ 0 \\ 0 \end{bmatrix} q_0(t), \quad x(0) = 0, \quad t \geq 0, \tag{69}$$

where $x(t) = [x_1(t)\ x_2(t)\ x_3(t)]^\mathsf{T}$ is the system state, $\delta(t, x(t))$ represents an uncertainty of the form given in (6) with,

$$W_0(t) = [\sin(0.25t), -0.25, 0.5, 0.5]^\mathsf{T}, t \geq 0, \tag{70}$$

$$\sigma_0(x(t)) = [x_1(t), x_2(t), x_1(t)x_2(t), x_3(t)]^\mathsf{T}, t \geq 0, \tag{71}$$

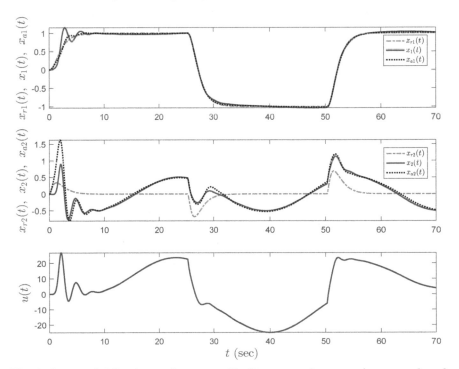

Fig. 6. Command following performance with the proposed command governor-based adaptive controller.

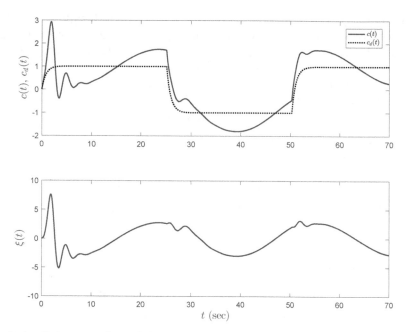

Fig. 7. Applied command signal $c(t)$, $t \geq 0$, and the command governor signal $\xi(t)$, $t \geq 0$.

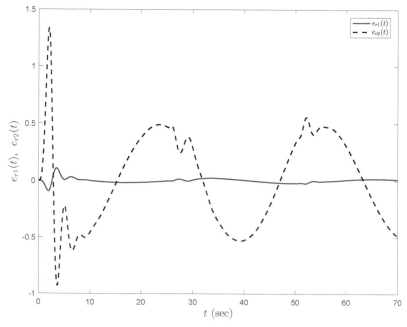

Fig. 8. The evolution of the error signal between the auxiliary and the reference system $e_r(t)$, $t \geq 0$.

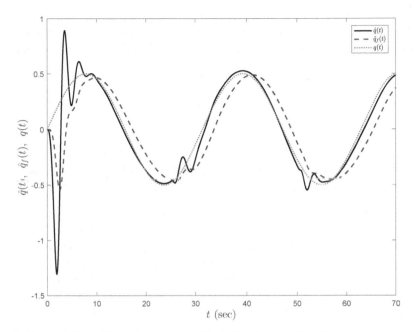

Fig. 9. The evolution of the unknown unmatched disturbance estimation.

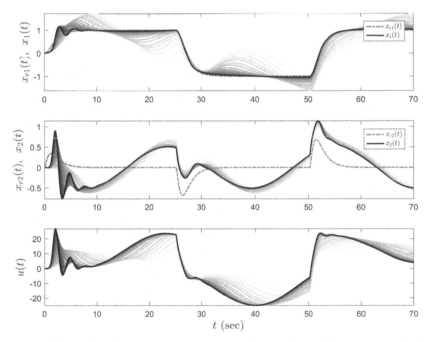

Fig. 10. The effect of increasing the design parameter $\Gamma = \Gamma_0 = \Gamma_1$ from 0.05 to 10 (light gray to black) on the system performance.

$q_0(t) = 0.5 \sin(0.1t)$, $t \geq 0$, represents the unmatched disturbance, and $\Lambda = 0.75$ represents an uncertain control effectiveness matrix. Linear quadratic regulator theory is used to design the nominal feedback gain matrix as:

$$K_1 = [5.7, 9.9, 6.1], \tag{72}$$

and we pick $K_2 = 3.7$. Now, in order to satisfy the condition rank($[B, D]$) = 3, we rewrite (69) equivalently as:

$$\dot{x}(t) = \begin{bmatrix} 0 & 1 & 0 \\ 0 & 0 & 1 \\ 2 & 3 & 1 \end{bmatrix} x(t) + \begin{bmatrix} 0 \\ 0 \\ 1 \end{bmatrix} \Big(\Lambda u(t) + \delta(t, x(t)) \Big) + \begin{bmatrix} 1 & 0 \\ 0 & 1 \\ 0 & 0 \end{bmatrix} \begin{bmatrix} q_0(t) \\ 0 \end{bmatrix}, \quad x(0) = 0, \quad t \geq 0. \tag{73}$$

For the adaptive controller in Section 4, we set the projection norm bound imposed on each element of the parameter estimate to $\hat{W}_{max} = 30$ and $\hat{q}_{max} = 5$ and the learning rates to $\gamma_q = 2$ and $\gamma_W = 10$ and we use $R = I$ to calculate P from (18) for the resulting A_r matrix. Figs. 12 and 13 show the closed-loop dynamical system performance with the standard adaptive controller in Section 4. One can see from this figure that the standard adaptive controller cannot compensate the effects of the unmatched

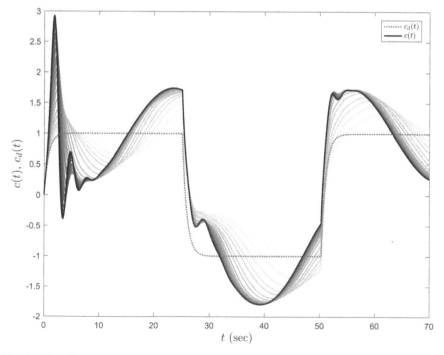

Fig. 11. The effect of increasing the design parameter $\Gamma = \Gamma_0 = \Gamma_1$ from 0.05 to 10 (light gray to black) on the modified command signal using the command governor signal.

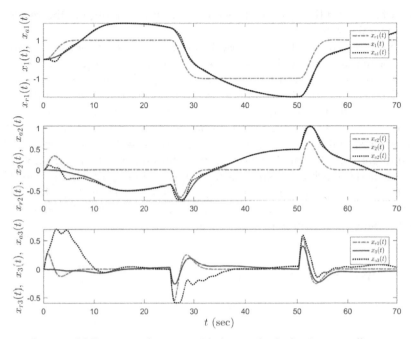

Fig. 12. Command following performance with the standard adaptive controller.

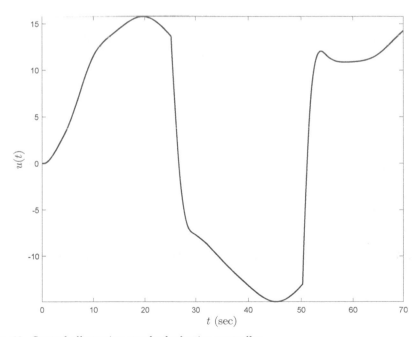

Fig. 13. Control effort using standard adaptive controller.

disturbance and the systems trajectories do not converge to the reference system trajectory.

Next, we apply the proposed command governor-based adaptive control architecture, with $\Gamma_0 = \Gamma_1 = \Gamma_2 = 10$ and the filter gain in (34) set to $\Gamma_f = 0.5$. It can be seen in Fig. 14 that desired performance is obtained and the first component of the state vector converges to a close vicinity of the reference state. Fig. 15 clearly shows the role of the command governor signal to modify the command signal such that the first component of the error signal $e_{r1}(t)$, $t \geq 0$, gets arbitrarily close to zero by tuning the design gains Γ_0, Γ_1 and Γ_2 as one can see in Fig. 16. The evolution of the unmatched disturbance estimation is depicted in Fig. 17. Finally, the effect of the design parameter $\Gamma = \Gamma_0 = \Gamma_1 = \Gamma_2$ can be seen in Figs. 18 and 19, which makes it clear that a larger value of Γ, leads to a better tracking performance of output signal of the reference system.

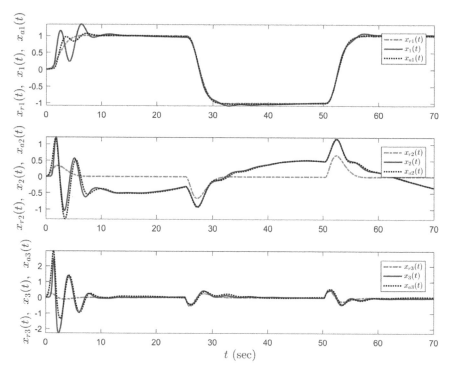

Fig. 14. Command following performance with the proposed command governor-based adaptive controller.

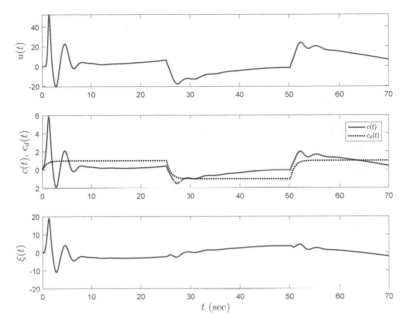

Fig. 15. The control effort $u(t)$, $t \geq 0$, the applied command signal $c(t)$, $t \geq 0$, and the command governor signal $\xi(t)$, $t \geq 0$.

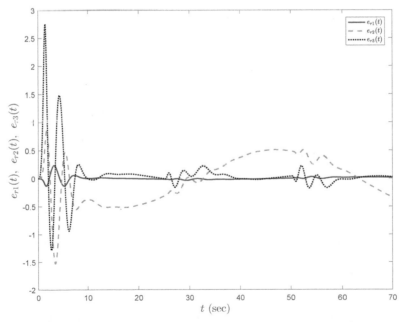

Fig. 16. The evolution of the error signal between the auxiliary and the reference system $e_r(t)$, $t \geq 0$.

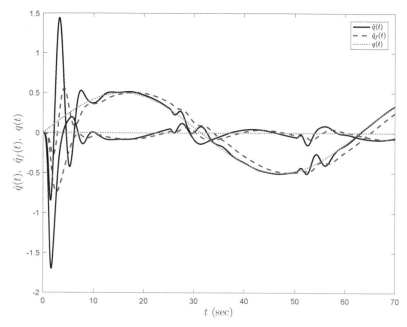

Fig. 17. The evolution of the unknown unmatched disturbance estimation.

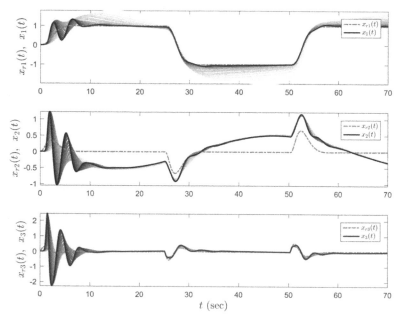

Fig. 18. The effect of increasing the design parameter $\Gamma = \Gamma_0 = \Gamma_1 = \Gamma_2$ from 0.05 to 10 (light gray to black) on the system performance.

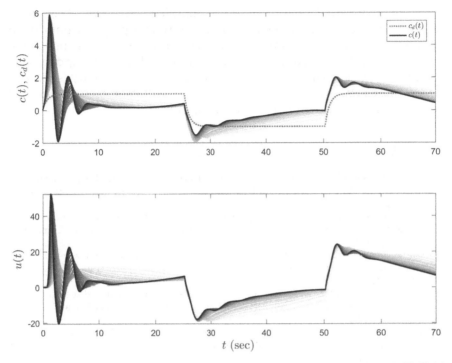

Fig. 19. The effect of increasing the design parameter $1 - 1_0 - 1'_1 - 1'_2$ from 0.05 to 10 (light gray to black) on the control effort and the modified command signal using the command governor signal.

7. CONCLUSION

A challenge in the design of model reference adaptive control architecture is to cope with the effect of unmatched disturbances while dealing with matched uncertainties. To this end, we proposed a two-level design framework based on a command governor architecture to suppress the effect of matched uncertainties and unmatched disturbances and achieve a close tracking of the output of the reference system. In particular, an auxiliary state dynamics was first designed to allow for the estimation of both matched uncertainties and unmatched disturbances. We then proposed a command governor architecture through a backstepping procedure to modify the command signal of the desired reference system such that the system output error signal can be made arbitrarily small by tuning the constant design parameters. Two numerical examples demonstrated the efficacy of our two-level design framework.

Considering numerous agriculture applications when the nature of the environmental disturbances are unmatched, the proposed command governor-based model reference adaptive control framework of this paper has a high potential to guarantee the completion of given tasks (e.g., autonomous seeding, harvesting, and/or row cropping via unmanned ground vehicles, or farm imaging and monitoring via unmanned aerial vehicles) with high accuracy. Future research will focus on applications of the proposed framework to real-world unmanned vehicles as well as on extensions to the dynamical systems with not only unmatched disturbances but also unmatched uncertainties.

REFERENCES

Boskovic, J. D., and Han, Z. 2009. Certainty equivalence adaptive control of plants with unmatched uncertainty using state feedback. IEEE Transactions on Automatic Control. 54(8): pp. 1918–1924.

Cao, C. and Hovakimyan, N. 2008. $\mathcal{L}_1$ adaptive controller for multi-input multi-output systems in the presence of unmatched disturbances. American Control Conference. IEEE. pp. 4105–4110.

Cariou, C., Lenain, R., Thuilot, B., and Berducat, M. 2009. Automatic guidance of a four-wheel-steering mobile robot for accurate field operations. Journal of Field Robotics. 26(6-7): pp. 504–518.

Che, J., and Cao, C. 2012. $\mathcal{L}_1$ adaptive control of system with unmatched disturbance by using eigenvalue assignment method. IEEE Conference on Decision and Control. pp. 4823–4828.

Chellaboina, V., Haddad, W. M., Bernstein, D. S., and Wilson, D. A. 2000. Induced convolution operator norms of linear dynamical systems. Mathematics of Control, Signals and Systems. pp. 216–239.

Fravolini, M. L., and Campa, G. 2011. Integrated design of a linear/neuro-adaptive controller in the presence of norm-bounded uncertainties. International Journal of Control. 84(10): pp. 1664–1677.

Haddad, W. M., and Chellaboina, V. 2008. Nonlinear dynamical systems and control: A Lyapunov-based approach. Princeton University Press.

Heise, C. D., and Holzapfel, F. 2015. Uniform ultimate boundedness of a model reference adaptive controller in the presence of unmatched parametric uncertainties. Automation, Robotics and Applications (ICARA), 2015 6th International Conference on, 2015. pp. 149–154.

Ioannou, P. A., and Sun, J. 2012. Robust adaptive control. Courier Corporation.

Kristic, M., Kanellakopoulos, I., and Kokotovic, P. 1995. Nonlinear and Adaptive Control Design. Adaptive and Learning systems for signal processing, communications and control. John Wiley and Sons.

Lavretsky, E., and Wise, K. 2012. Robust and adaptive control with aerospace applications. Springer Science & Business Media.

Leman, T. J., Xargay, E., Dullerud, G., Hovakimyan, N., and Wendel, T. 2010. $\mathcal{L}_1$ adaptive control augmentation system for the X-48B aircraft. University of Illinois.

Lenain, R., Thuilot, B., Cariou, C., and Martinet, P. 2003. Adaptive control for car like vehicles guidance relying on rtk gps: Rejection of sliding effects in agricultural applications. IEEE International Conference on Robotics and Automation. 1: pp. 115–120.

Li, Z., and Hovakimyan, N. 2012. $\mathcal{L}_1$ adaptive controller for mimo systems with unmatched uncertainties using modified piecewise constant adaptation law. IEEE Conference on Decision and Control. pp. 7303–7308.

Mäkynen, J., Saari, H., Holmlund, C., Mannila, R., and Antila, T. 2012. Multi-and hyperspectral UAV imaging system for forest and agriculture applications, Proc. SPIE 8374, Next-Generation Spectroscopic Technologies V, 837409, Baltimore, MD.

McRuer, D. T., Graham, D., and Ashkenas, I. 2014. Aircraft dynamics and automatic control. Princeton University Press.

Narendra, K. S., and Annaswamy, A. M. 2012. Stable adaptive systems. Courier Corporation.

Pomet, J. -B., and Praly, L. 1992. Adaptive nonlinear regulation: Estimation from the lyapunov equation. IEEE Transactions on Automatic Control. 37(6): 729–740.

Primicerio, J., Di Gennaro, S. F., Fiorillo, E., Genesio, L., Lugato, E., Matese, A., and Vaccari, F. P. 2012. A flexible unmanned aerial vehicle for precision agriculture. Precision Agriculture. 13(4): pp. 517–523.

Saari, H., Pellikka, I., Pesonen, L., Tuominen, S., Heikkilä, J., Holmlund, C., Mäkynen, J., Ojala, K., and Antila, T. 2011. Unmanned Aerial Vehicle (UAV) operated spectral camera system for forest and agriculture applications. Proc. SPIE 8174, Remote Sensing for Agriculture, Ecosystems, and Hydrology XIII, 81740H, Prague, Czech Republic.

Stengel, R. F. 2015. Flight dynamics. Princeton University Press.

Stepanyan, V., and Krishnakumar, K. 2012. Certainty equivalence m-mrac for systems with un-matched uncertainties. IEEE Conference on Decision and Control. pp. 4152–4157.

Stepanyan, V., and Krishnakumar, K. 2015. State and output feedback certainty equivalence m-mrac for systems with unmatched uncertainties. Asian Journal of Control. 17(6): pp. 2041–2054.

Tokekar, P., Vander Hook, J., Mulla, D., and Isler, V. 2016. Sensor planning for a symbiotic uav and ugv system for precision agriculture. IEEE Transactions on Robotics. 32(6): pp. 1498–1511.

Xargay, E., Hovakimyan, N., and Cao, C. 2010. $\mathcal{L}_1$ adaptive controller for multi-input multi output systems in the presence of nonlinear unmatched uncertainties. American Control Conference. pp. 874–879.

Yayla, M., and Turker Kutay, A. 2016. Adaptive control algorithm for linear systems with matched and unmatched uncertainties. IEEE Conference on Decision and Control. pp. 2975–2980.

Yucelen, T., and Haddad, W. M. 2012. A robust adaptive control architecture for disturbance rejection and uncertainty suppression with $\mathcal{L}_\infty$ transient and steady-state performance guarantees. International Journal of Adaptive Control and Signal Processing, pp. 1024–1055.

Index

Milton Keynes UK
Ingram Content Group UK Ltd.
UKHW040100071024
449327UK00019B/687